T0252028

RELIABILITY ANALYSIS

WITH
MINITAB®

RELIABILITY ANALYSIS

WITH

MINITAB®

Kishore K. **Pochampally**
Surendra M. **Gupta**

CRC Press
Taylor & Francis Group
Boca Raton London New York

CRC Press is an imprint of the
Taylor & Francis Group, an **informa** business

CRC Press
Taylor & Francis Group
6000 Broken Sound Parkway NW, Suite 300
Boca Raton, FL 33487-2742

Printed on acid-free paper
Version Date: 20151103

International Standard Book Number-13: 978-1-4987-3758-6 (Hardback)

Library of Congress Cataloging-in-Publication Data

Names: Pochampally, Kishore K., author. | Gupta, Surendra M.
Title: Reliability analysis with Minitab / authors, Kishore K. Pochampally
and Surendra M. Gupta.
Description: Boca Raton : Taylor & Francis, 2016. | Includes bibliographical
references and index.
Identifiers: LCCN 2015042089 | ISBN 9781498737586 (hard cover)
Subjects: LCSH: Reliability (Engineering)--Mathematics. | Minitab.
Classification: LCC TS173 .P59 2016 | DDC 620/.00452--dc23
LC record available at http://lccn.loc.gov/2015042089

Visit the Taylor & Francis Web site at
http://www.taylorandfrancis.com

and the CRC Press Web site at
http://www.crcpress.com

To my greatest sources of joy: Sai Pochampally, Seema Dasgupta, and Buddy Pochampally.

Kishore K. Pochampally

To Sharda Gupta, for her unconditional support.

Surendra M. Gupta

Contents

Section I   Introduction

Section II   Case Studies

Preface

Reliability is the probability that a product will perform as intended for a specific duration under normal conditions. Improper or ineffective reliability analysis during the product's design phase leads to product failures which, in turn, result in excessive costs and even fatalities. For example, the return of a car to the manufacturer within the warranty period, an accident due to a brake failure, and a dissatisfied customer due to a poorly functioning product may all be prevented by effective reliability analysis.

Since reliability, by definition, is a probability, and since a probability is estimated using knowledge of statistics, it is necessary that reliability engineers are well versed in statistical concepts and applications. Minitab® is the world's leading statistical software used for reliability analysis and quality improvement. It is user-friendly, has hundreds of sample data sets, and can help in performing the very basic to advanced statistical analyses. It has a set of reliability analysis tools that can perform tasks ranging from checking the distribution fit of lifetime data to estimation of warranty costs of a product.

This book illustrates the application of Minitab for reliability analysis using over a dozen different case studies. Over 200 screenshots are used to demonstrate concepts such as parametric reliability analysis, nonparametric reliability analysis, warranty analysis, accelerated life testing, reliability test plans, and probit analysis.

Additional material is available from the CRC website: http://www.crc-press.com/product/isbn/9781498737586.

List of Figures

List of Tables

Acknowledgments

We would like to thank Bill Gillett (Southern New Hampshire University), Mary Viola (Tufts University), Elif Kongar (University of Bridgeport), Hanchen Huang (Northeastern University), Cindy Renee Carelli (Taylor & Francis), and Laurie Oknowsky (Taylor & Francis) for their help and support during this project.

Kishore K. Pochampally
Surendra M. Gupta

Authors

Kishore K. Pochampally is a professor of quantitative studies, operations, and project management at Southern New Hampshire University in Manchester. He also conducts corporate workshops in lean six sigma quality and project management. His prior academic experience is as a postdoctoral fellow at Massachusetts Institute of Technology in Cambridge. He holds graduate degrees (MS and PhD) in industrial engineering from Northeastern University in Boston, Massachusetts, and an undergraduate degree (BE) in mechanical engineering from the National Institute of Technology in India. His research interests are in the areas of six sigma quality management and quantitative decision-making. He has authored a number of technical papers for international journals and conference proceedings, and his research work has been cited by other researchers on six continents. He has published two books: *Strategic Planning Models for Reverse and Closed-Loop Supply Chains* and *Six Sigma Case Studies with Minitab®*. He has been nominated twice for the teaching excellence award at Southern New Hampshire University. He is also a Six Sigma Black Belt (American Society for Quality) and a Project Management Professional.

Surendra M. Gupta is a professor of mechanical and industrial engineering and the director of the laboratory for responsible manufacturing at Northeastern University. He received his BE in electronics engineering from Birla Institute of Technology and Science, MBA from Bryant University, and MSIE and PhD in industrial engineering from Purdue University. He is a registered professional engineer in the state of Massachusetts. Dr. Gupta's research interests are in the areas of production/manufacturing systems and operations research. He is mostly interested in environmentally conscious manufacturing, reverse and closed-loop supply chains, disassembly modeling, and remanufacturing. He has authored or coauthored well over 500 technical papers, published in books, journals, and international conference proceedings. His publications have been cited by thousands of researchers all over the world in journals, proceedings, books, and dissertations. He has traveled to all seven continents, viz., Africa, Antarctica, Asia, Australia, Europe, North America, and South America, and presented his work at international conferences on six continents. Dr. Gupta has taught over 100 courses in areas such as operations research, inventory theory, queuing theory, engineering economy, supply chain management, and production planning and control. Among the many accolades received, he is the recipient of the outstanding research award, and the outstanding industrial engineering professor award (in recognition of teaching excellence) from Northeastern University, as well as the national outstanding doctoral dissertation advisor award.

Section I

Introduction

1

Fundamental Concepts in Reliability Analysis

Reliability of a part or product (which will be referred to as a *unit* hereafter) can be defined as the probability of the unit performing successfully for a specific period of time. For example, if the reliability of a car giving at least 40 miles per gallon for 2 years (as intended) is 90%, it means that there is a 90% probability that the car performs as intended.

Since a *probability* is estimated using *statistics*, a reliability engineer must be well versed in both probability theory and statistical analysis. To this end, this chapter introduces some fundamental concepts and terms used in reliability analysis.

The concepts of *exact failure time, right-censored failure time, left-censored failure time*, and *interval-censored failure time* are explained in Section 1.1. Section 1.2 illustrates the *probability density function*, the *survival function*, and the *hazard rate function*. Finally, an outline of the book is presented in Section 1.3.

1.1 Types of Failure Time

An *exact failure time* of a unit means that the test unit has failed exactly at that time. For example, if a product has failed exactly at the end of, say, 50 hours, the failure time is recorded as 50 hours exact.

A *right-censored failure time* means that the test unit has not failed by that time. For example, if a product has not failed in, say, 100 hours, and if the test duration has ended, the failure time is recorded as 100 hours right-censored. This product has not failed in 100 hours, but is expected to fail anytime later.

A *left-censored failure time* means that the test unit failed by that time, but we do not know exactly when. For example, if a product has failed within, say, 100 hours, and if we do not know exactly when, the failure time is recorded as 100 hours left-censored.

An *interval-censored failure time* means that the test unit failed within a time *interval*, but we do not know exactly when. For example, if a product has failed after 100 hours and by the end of 200 hours, but we do not know exactly when, the failure time is recorded as 100 hours–200 hours interval-censored.

An *arbitrarily censored failure time data set* may consist of exact failure times, right-censored failure times, left-censored failure times, and/or interval-censored failure times.

1.2 Probability Functions in Reliability

Table 1.1 shows the failure time data from a test conducted on 250 randomly selected products.

Notice from the table that 17 products have failed by 10,000 hours (left-censored), 56 products survived past 90,000 hours (right-censored), and the remaining products failed in different intervals of time (e.g., 24 products survived past 10,000 hours but failed within 20,000 hours). There are no exact-censored failure times in this table, but if, say, 10 products failed exactly at the end of, say, 13,000 hours, we would have entered 13,000 hours in both Start and End columns, and 10 products in the Frequency column.

A *probability density function* of a failure time shows the relative probability of the failure time to be a given value. Figure 1.1 illustrates the probability density function for the data given in Table 1.1. In later chapters, it will be illustrated as the statistical distribution that is the best fit for a given set of failure times. For Table 1.1, however, it is assumed that the failure times follow a normal distribution. It is evident from Figure 1.1 that the expected (average) failure time for the product is slightly over 50,000 hours, and that the relative probability of a product failing at a certain time decreases as that time gets farther from the expected (average) failure time.

A *survival function* of a failure time shows the probability of a unit surviving past that failure time. Figure 1.2 shows the survival function for the data

TABLE 1.1

Failure Time Data

Start	End	Frequency
*	10,000	17
10,000	20,000	24
20,000	30,000	32
30,000	40,000	21
40,000	50,000	17
50,000	60,000	34
60,000	70,000	23
70,000	80,000	11
80,000	90,000	15
90,000	*	56

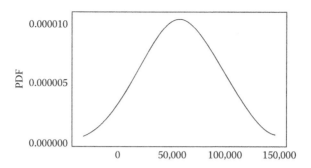

FIGURE 1.1
Probability density function.

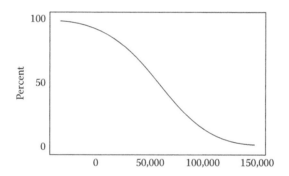

FIGURE 1.2
Survival function.

given in Table 1.1. It is clear from Figure 1.2 that about 50% of the products are expected to survive a little over 50,000 hours, and about 15% of the products are expected to survive past 100,000 hours.

A *hazard rate function* of a failure time shows the probability of a unit failing at that time, given that it has survived until then. It is the ratio of the probability density function value at that time to the survival function value at that time. Figure 1.3 illustrates the hazard rate function for the data given in Table 1.1. It is clear from Figure 1.3 that the longer this product survives, the greater its probability of failure at that instant.

Readers interested in the mathematical formulas of probability density function, survival function, and hazard rate function are encouraged to review the books listed in the Bibliography.

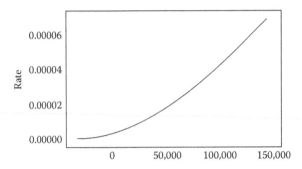

FIGURE 1.3
Hazard rate function.

1.3 Book Outline

This book is divided into two parts: Section I is the Introduction and Section II is the Case Studies. Section I consists of Chapters 1 and 2, and Section II consists of Chapters 3 through 12.

- Chapter 1 explains some fundamental concepts used in reliability analysis.
- Chapter 2 illustrates the shapes of the most commonly used distributions in reliability analysis.
- Chapter 3 presents a case study that illustrates how to perform reliability analysis of a product with right-censored and exact failure time data.
- Chapter 4 presents a case study that illustrates how to perform reliability analysis of a product with arbitrarily censored failure time data.
- Chapter 5 presents a case study that illustrates how to perform non-parametric reliability analysis of a product.
- Chapter 6 presents a case study that illustrates how to predict the amount of money that is needed to cover the warranty costs for products in a specific period of time in the future.
- Chapter 7 presents a case study that illustrates how to analyze the results from accelerated life testing on two different products.
- Chapter 8 presents a case study that illustrates how to determine the reliability test sample size when the test time and the number of failures are constrained.

- Chapter 9 presents a case study that illustrates how to determine the testing time when test sample size and the number of failures are constrained.
- Chapter 10 presents a case study that illustrates how to compare the reliabilities of parts from different vendors.
- Chapter 11 presents a case study that illustrates how to test whether the reliability of a product depends on certain factors.
- Chapter 12 presents a case study that illustrates how to predict the stress level at which a product will fail after a certain test period.

2

Commonly Used Statistical
Distributions in Reliability Analysis

Before performing reliability analysis of a unit, it is important to collect failure time data from a test sample and check which statistical distribution best fits the data. Minitab® considers the following most commonly used distributions (probability density functions) in reliability analysis, whose shapes are illustrated in Sections 2.1 through 2.11 of this chapter:

- Smallest extreme value
- Weibull
- Three-parameter Weibull
- Exponential
- Two-parameter exponential
- Normal
- Lognormal
- Three-parameter lognormal
- Logistic
- Loglogistic
- Three-parameter loglogistic

Section 2.12 gives a concluding remark.

Readers interested in the mathematical formulas of the above distributions are encouraged to review the handbook listed in the Bibliography.

2.1 Smallest Extreme Value

A smallest extreme value distribution function is defined by two parameters: location (of which the mean is a function) and scale (the larger the scale, the wider the graph). Figure 2.1 illustrates the function with a location of 3.0 and a scale of 1.0. It is evident from Figures 2.2 and 2.3 how the graph changes when the scale is changed to 2.0 and the location is changed to 4.0, respectively.

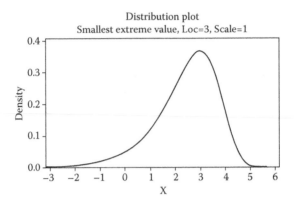

FIGURE 2.1
Smallest extreme value with location 3 and scale 1.

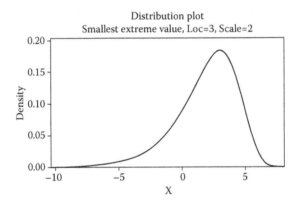

FIGURE 2.2
Smallest extreme value with location 3 and scale 2.

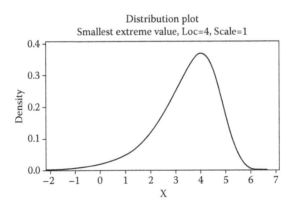

FIGURE 2.3
Smallest extreme value with location 4 and scale 1.

2.2 Weibull

A Weibull distribution function is defined by two parameters: shape (shape of the graph) and scale (the larger the scale, the wider the graph). Figure 2.4 illustrates the function with a shape of 2.0 and a scale of 3.0. It is evident from Figures 2.5 and 2.6 how the graph changes when the shape is changed to 3.0 and the scale is changed to 4.0, respectively.

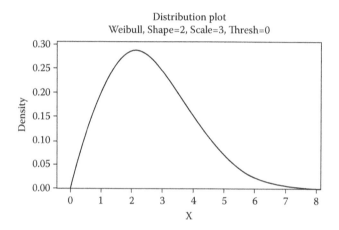

FIGURE 2.4
Weibull with shape 2 and scale 3.

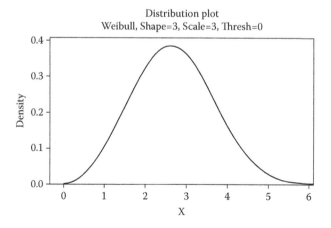

FIGURE 2.5
Weibull with shape 3 and scale 3.

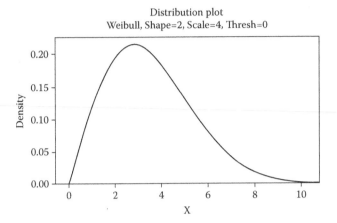

FIGURE 2.6
Weibull with shape 2 and scale 4.

2.3 Three-Parameter Weibull

Compared with the regular Weibull distribution (see Section 2.2), a three-parameter Weibull distribution has an additional parameter called the threshold (minimum value). Figure 2.7 illustrates how the graph changes from the one in Figure 2.4 when the threshold is changed from 0.0 (default value) to 2.0.

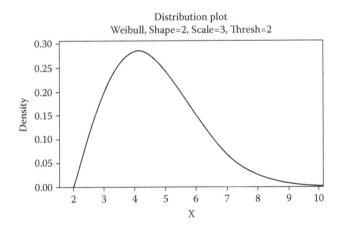

FIGURE 2.7
Three-parameter Weibull with shape 2, scale 3, and threshold 2.

2.4 Exponential

An exponential distribution function is defined by one parameter: scale (the larger the scale, the wider the graph). Figures 2.8 and 2.9 illustrate the function with scales of 1.0 and 2.0, respectively.

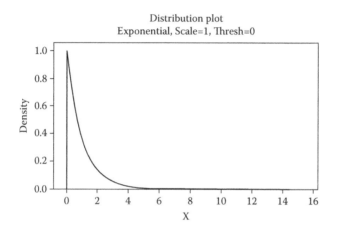

FIGURE 2.8
Exponential with scale 1.

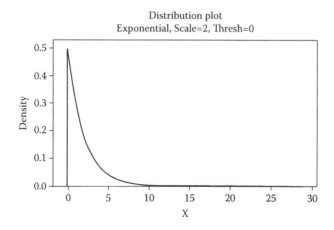

FIGURE 2.9
Exponential with scale 2.

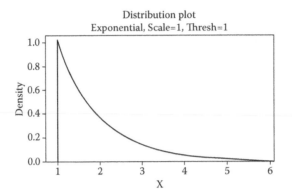

FIGURE 2.10
Two-parameter exponential with scale 1 and threshold 1.

2.5 Two-Parameter Exponential

Compared with the regular exponential distribution (see Section 2.4), a two-parameter exponential distribution has an additional parameter called the threshold (minimum value). Figure 2.10 illustrates how the graph changes from the one in Figure 2.8 when the threshold is changed from 0.0 (default value) to 1.0.

2.6 Normal

A normal distribution function is defined by two parameters: mean and standard deviation. Figure 2.11 illustrates the function with a mean of 0.0

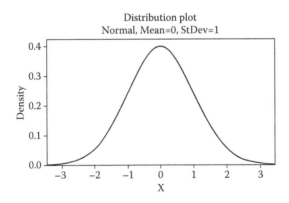

FIGURE 2.11
Normal with mean 0 and standard deviation 1.

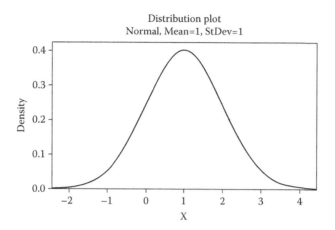

FIGURE 2.12
Normal with mean 1 and standard deviation 1.

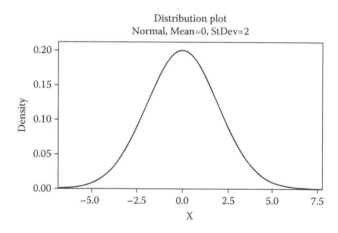

FIGURE 2.13
Normal with mean 0 and standard deviation 2.

and a standard deviation of 1.0. It is evident from Figures 2.12 and 2.13 how the graph changes when the mean is changed to 1.0 and the standard deviation is changed to 2.0, respectively.

2.7 Lognormal

A lognormal distribution function is defined by two parameters: location (of which the mean is a function) and scale (the larger the scale, the wider the

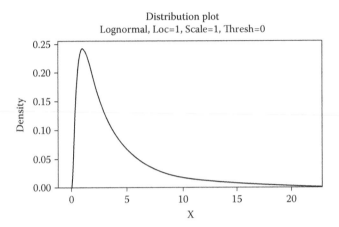

FIGURE 2.14
Lognormal with location 1 and scale 1.

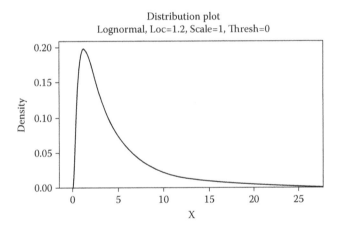

FIGURE 2.15
Lognormal with location 1.2 and scale 1.

graph). Figure 2.14 illustrates the function with a location of 1.0 and a scale of 1.0. It is evident from Figures 2.15 and 2.16 how the graph changes when the location is changed to 1.2 and the scale is changed to 1.3, respectively.

2.8 Three-Parameter Lognormal

Compared with the regular lognormal distribution (see Section 2.7), a three-parameter lognormal distribution has an additional parameter called the

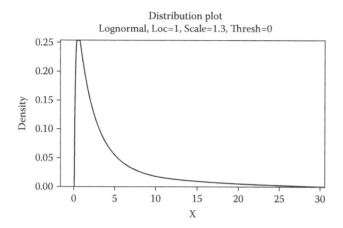

FIGURE 2.16
Lognormal with location 1 and scale 1.3.

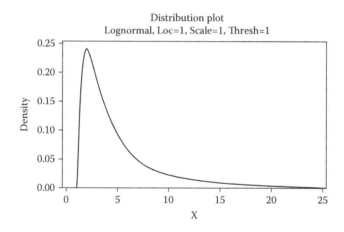

FIGURE 2.17
Three-parameter lognormal with location 1, scale 1, and threshold 1.

threshold (minimum value). Figure 2.17 illustrates how the graph changes from the one in Figure 2.14 when the threshold is changed from 0.0 (default value) to 1.0.

2.9 Logistic

A logistic distribution function is defined by two parameters: location (of which the mean is a function) and scale (the larger the scale, the wider the

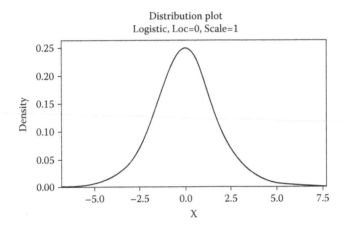

FIGURE 2.18
Logistic with location 0 and scale 1.

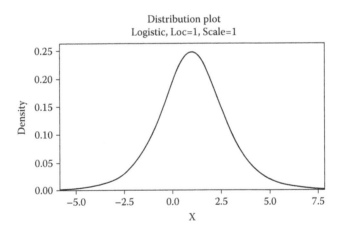

FIGURE 2.19
Logistic with location 1 and scale 1.

graph). Figure 2.18 illustrates the function with a location of 0.0 and a scale of 1.0. It is evident from Figures 2.19 and 2.20 how the graph changes when the location is changed to 1.0 and the scale is changed to 3.0, respectively.

2.10 Loglogistic

A loglogistic distribution function is defined by two parameters: location (of which the mean is a function) and scale (the larger the scale, the wider the

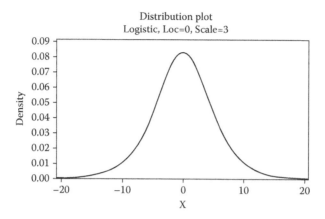

FIGURE 2.20
Logistic with location 0 and scale 3.

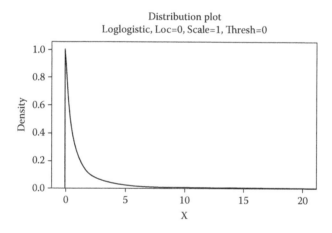

FIGURE 2.21
Loglogistic with location 0 and scale 1.

graph). Figure 2.21 illustrates the function with a location of 0.0 and a scale of 1.0. It is evident from Figures 2.22 and 2.23 how the graph changes when the location is changed to 1.0 and the scale is changed to 1.1, respectively.

2.11 Three-Parameter Loglogistic

Compared with the regular loglogistic distribution (see Section 2.10), a three-parameter loglogistic distribution has an additional parameter called the threshold (minimum value). Figure 2.24 illustrates how the

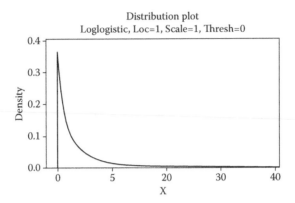

FIGURE 2.22
Loglogistic with location 1 and scale 1.

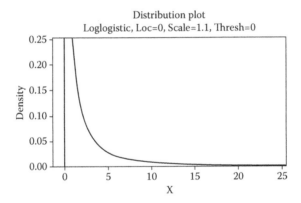

FIGURE 2.23
Loglogistic with location 0 and scale 1.1.

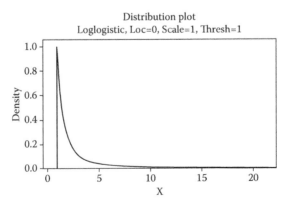

FIGURE 2.24
Three-parameter loglogistic with location 0, scale 1, and threshold 1.

graph changes from the one in Figure 2.21 when the threshold is changed from 0.0 (default value) to 1.0.

2.12 Conclusion

This chapter illustrated the shapes of the most commonly used statistical distributions in reliability analysis.

Section II

Case Studies

3

Reliability Analysis with Right-Censored and Exact Failure Times

Recall from Chapter 1 that a right-censored failure time means that the unit that is being tested has not failed by that time. For example, if a unit has not failed in, say, 100 hours, and if the test duration has ended, the failure time is recorded as 100 hours (right-censored). This unit has not failed in 100 hours, but is expected to fail anytime later. Recall also that an exact failure time means that the unit has failed exactly at that time. For example, if a unit has failed exactly at the end of, say, 50 hours, the failure time is recorded as 50 hours (exact).

In the case study considered here, a reliability engineer at a ceiling fan manufacturing firm wishes to perform reliability analysis on a particular brand of fans. Specifically, he is interested in knowing the answers to the following questions:

- By what time will 15% of the fans fail?
- By what time will 92% of the fans fail?
- What percentage of fans will survive past 70 months?

Table 3.1 shows the failure time data from a test conducted on 50 randomly selected fans. In the "Censored" column of the table, 1 refers to exact failure and 0 refers to right-censored failure.

Section 3.1 illustrates the approach to identify the statistical distribution that best fits the failure time data in Table 3.1. Section 3.2 gives the approach to perform reliability analysis of the data. Finally, Section 3.3 gives some concluding remarks.

3.1 Determination of Statistical Distribution

Perform the following steps to determine the statistical distribution that best fits the failure time data:

Step 1: Open a new Minitab® project file and copy the data from Table 3.1, as shown in Figure 3.1.

TABLE 3.1

Failure Time Data

No.	Failure Time (months)	Censored
1	55	1
2	58	1
3	53	1
4	40	1
5	51	1
6	99	0
7	35	1
8	55	1
9	74	1
10	101	0
11	56	1
12	45	1
13	61	1
14	92	0
15	73	0
16	51	1
17	49	1
18	24	1
19	37	1
20	31	1
21	67	1
22	62	1
23	100	0
24	58	1
25	46	1
26	51	1
27	27	1
28	52	1
29	48	1
30	79	0
31	48	1
32	67	1
33	66	1
34	27	1
35	59	1
36	48	1
37	77	0
38	58	1
39	51	1
40	97	0
41	34	1

TABLE 3.1 (CONTINUED)

Failure Time Data

No.	Failure Time (months)	Censored
42	79	0
43	91	0
44	41	1
45	64	1
46	81	0
47	105	0
48	84	0
49	54	1
50	23	1

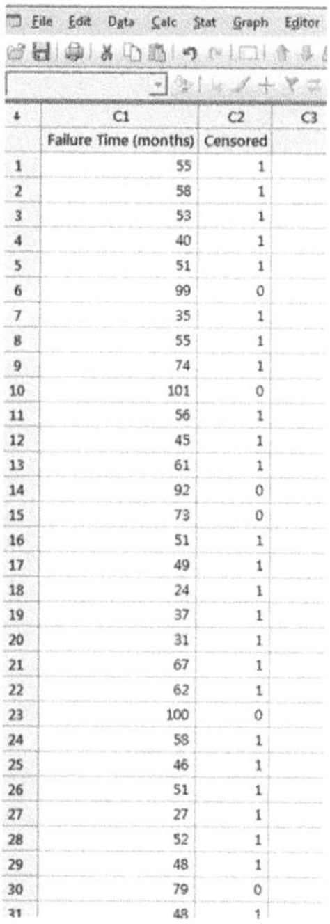

FIGURE 3.1
Step 1 in Section 3.1.

Step 2: As shown in Figure 3.2, go to *Stat* → *Reliability/Survival* → *Distribution Analysis (Right Censoring)* → *Distribution ID Plot*. The dialog box shown in Figure 3.3 opens.

Step 3: As shown in Figure 3.4, select the "Failure Time (months)" column for *Variables* and select the four default distributions: Weibull, Lognormal, Exponential, and Normal.

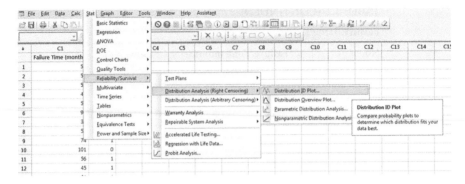

FIGURE 3.2
Step 2 commands in Section 3.1.

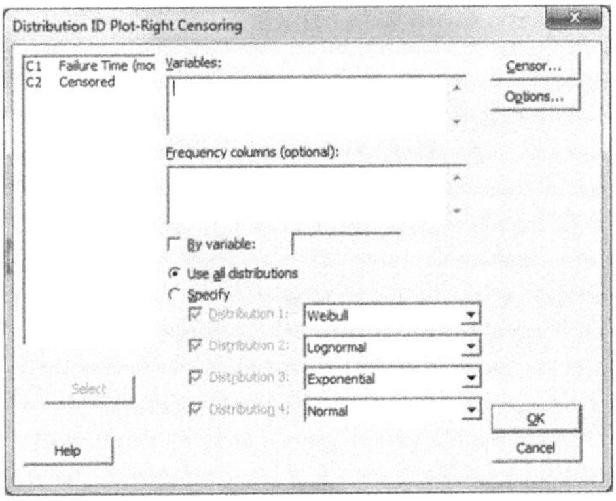

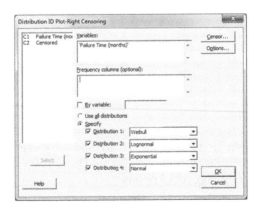

FIGURE 3.3
Step 2 dialog box in Section 3.1.

FIGURE 3.4
Step 3 in Section 3.1.

Step 4: Click on *Censor,* and the dialog box shown in Figure 3.5 opens.

Step 5: As shown in Figure 3.6, select the "Censored" column for *Use censoring columns* and enter "0" for *Censoring value.*

Step 6: Click on *OK,* and it will take you back to the dialog box shown in Figure 3.4.

Step 7: Click on *Options,* and the dialog box shown in Figure 3.7 opens.

Step 8: As shown in Figure 3.8, add "15" and "92" in the box for *Estimate percentiles for these percents.*

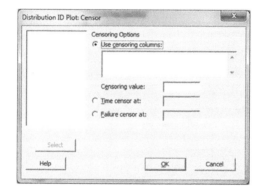

FIGURE 3.5
Step 4 in Section 3.1.

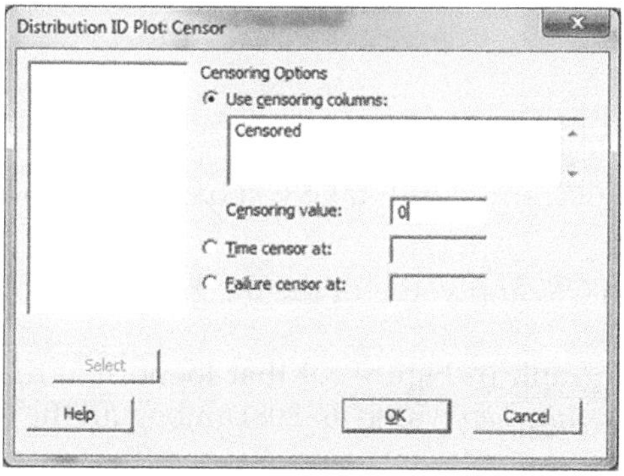

FIGURE 3.6
Step 5 in Section 3.1.

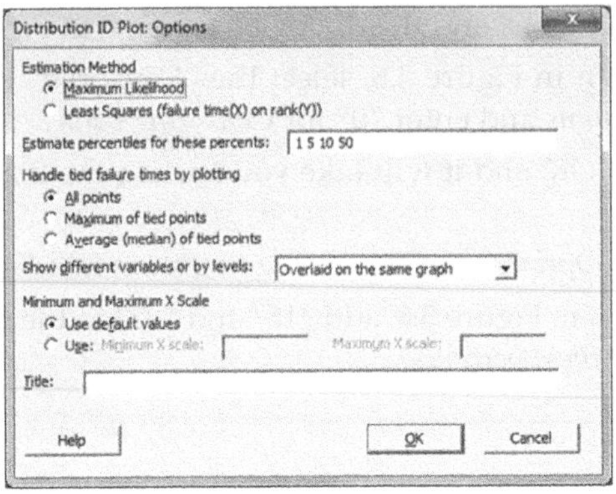

FIGURE 3.7
Step 7 in Section 3.1.

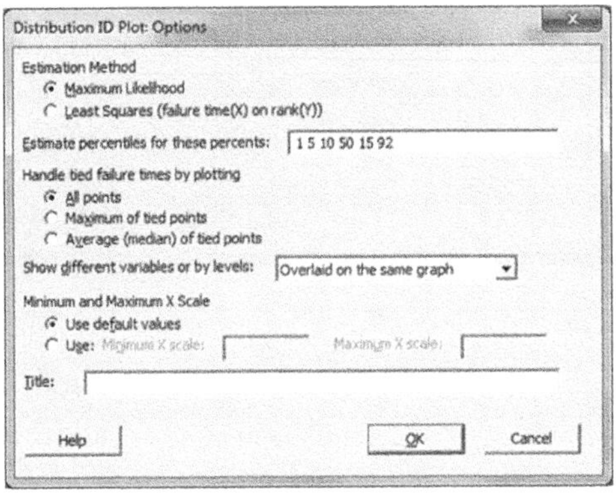

FIGURE 3.8
Step 8 in Section 3.1.

Step 9: Click on *OK*, and it will take you back to the dialog box shown in Figure 3.4.

Step 10: Click on *OK*, and you will see the graph shown in Figure 3.9.

Notice from the graph in Figure 3.9 that lognormal distribution has the lowest *Anderson–Darling (adj)* value (67.808) among all the four distributions.

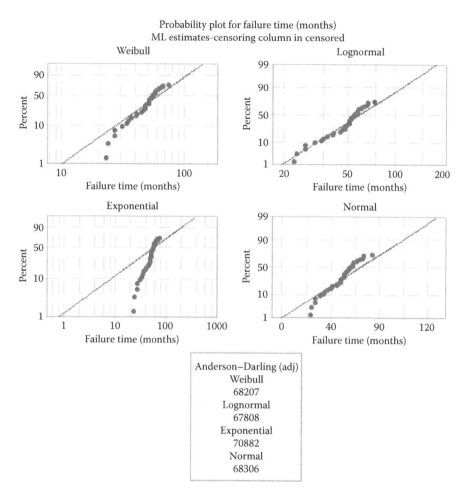

FIGURE 3.9
Step 10 in Section 3.1.

This is because lognormal distribution best fits the failure time data among all four distributions. Notice that the points are more close to the straight line for lognormal distribution than they are to the straight lines for the other three distributions.

In addition to the graph shown in Figure 3.9, you will see the following output in the session window:

```
Table of Percentiles

                                       Standard     95% Normal CI
Distribution   Percent   Percentile     Error     Lower      Upper
Weibull           1       10.1347      2.79369    5.90438    17.3960
Lognormal         1       19.3773      2.84108    14.5376    25.8283
Exponential       1        0.810546    0.133253   0.587274    1.11870
Normal            1       -0.393549    8.35837   -16.7757    15.9886

Weibull           5       20.4395      3.79672    14.2022    29.4161
Lognormal         5       26.9790      3.02887    21.6503    33.6193
Exponential       5        4.13673     0.680075    2.99724    5.70945
Normal            5       18.3551      6.39342    5.82418    30.8859

Weibull          10       27.8621      4.12241    20.8484    37.2354
Lognormal        10       32.1847      3.09621    26.6540    38.8629
Exponential      10        8.49718     1.39693    6.15656    11.7277
Normal           10       28.3499      5.47230    17.6244    39.0754

Weibull          15       33.5756      4.24476    26.2066    43.0166
Lognormal        15       36.2532      3.14831    30.5792    42.9800
Exponential      15       13.1069      2.15477    9.49652    18.0900
Normal           15       35.0933      4.93445    25.4220    44.7647

Weibull          50       62.6801      4.61776    54.2525    72.4168
Lognormal        50       59.9705      4.31092    52.0895    69.0439
Exponential      50       55.9014      9.19013    40.5029    77.1541
Normal           50       63.6066      4.06308    55.6431    71.5701

Weibull          92      109.349       9.77090    91.7816   130.279
Lognormal        92      118.653      14.2996     93.6902   150.266
Exponential      92      203.697      33.4875    147.587    281.139
Normal           92      102.262       6.83784    88.8597   115.664


Table of MTTF

                          Standard     95% Normal CI
Distribution     Mean      Error     Lower      Upper
Weibull        65.0252    4.6013    56.6043    74.699
Lognormal      67.4760    5.5497    57.4303    79.279
Exponential    80.6486   13.2586    58.4333   111.310
Normal         63.6066    4.0631    55.6431    71.570
```

From this output, it is evident that

- 1% of the fans are expected to fail in 19.3773 months.
- 5% of the fans are expected to fail in 26.9790 months.
- 10% of the fans are expected to fail in 32.1847 months.
- 15% of the fans are expected to fail in 36.2532 months.
- 50% of the fans are expected to fail in 59.9705 months.
- 92% of the fans are expected to fail in 118.653 months.
- On average, a fan is expected to fail in 67.4760 months (mean time to failure, MTTF).

3.2 Reliability Analysis

Perform the following steps to carry out the reliability analysis of the data:

Step 1: As shown in Figure 3.10, go to *Stat → Reliability/Survival → Distribution Analysis (Right Censoring) → Distribution Overview Plot*. The dialog box shown in Figure 3.11 opens.

Step 2: As shown in Figure 3.12, select the "Failure Time (months)" column for *Variables*, and "Lognormal" for *Parametric analysis (distribution)*.

Step 3: Click on *Censor*, and the dialog box shown in Figure 3.13 opens.

Step 4: As shown in Figure 3.14, select the "Censored" column for *Use censoring columns* and enter "0" for *Censoring value*.

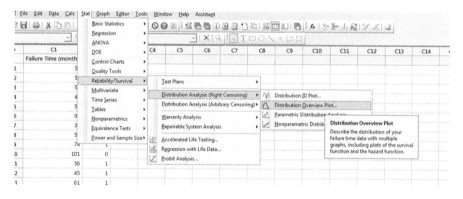

FIGURE 3.10
Step 1 commands in Section 3.2.

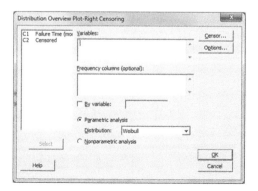

FIGURE 3.11
Step 1 dialog box in Section 3.2.

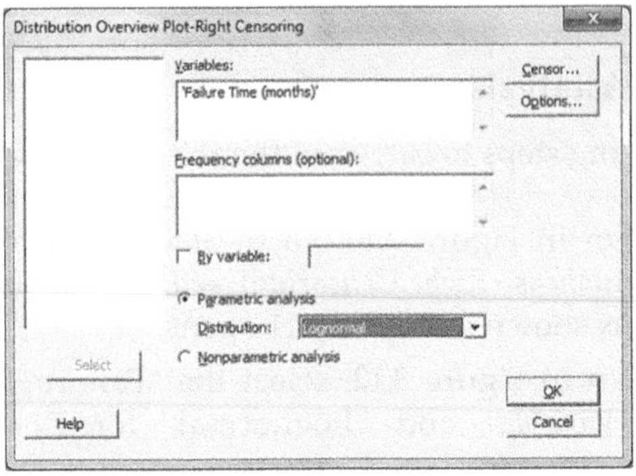

FIGURE 3.12
Step 2 in Section 3.2.

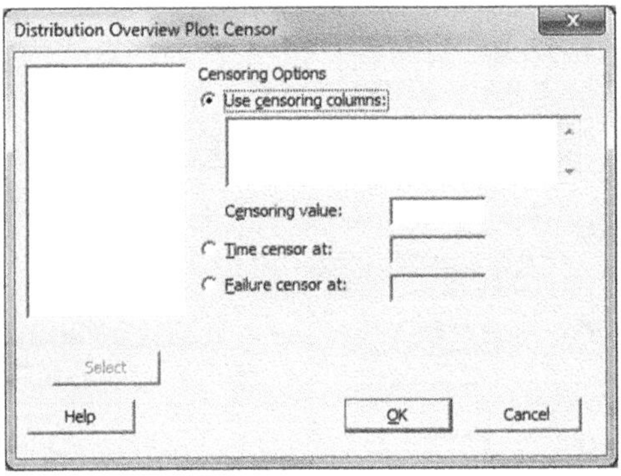

FIGURE 3.13
Step 3 in Section 3.2.

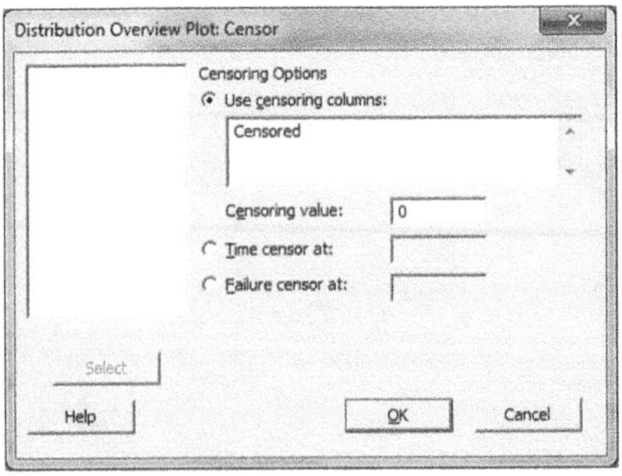

FIGURE 3.14
Step 4 in Section 3.2.

Step 5: Click on *OK*, and it will take you back to the dialog box shown in Figure 3.12.

Step 6: Click on *OK*, and you will see the graph shown in Figure 3.15. The graph displays the following for the failure time data: probability density function, distribution fit (lognormal), survival function, and hazard rate function. These graphs help the reliability engineer get a sense of the properties of the failure time data. For example, the hazard rate increases rapidly in the early stages of the product's life and then decreases slowly after 100 months.

Step 7: As shown in Figure 3.16, go to *Stat* → *Reliability/Survival* → *Distribution Analysis (Right Censoring)* → *Parametric Distribution Analysis*. The dialog box shown in Figure 3.17 opens.

Step 8: As shown in Figure 3.18, select the "Failure Time (months)" column for *Variables* and "Lognormal" for *Assumed distribution*.

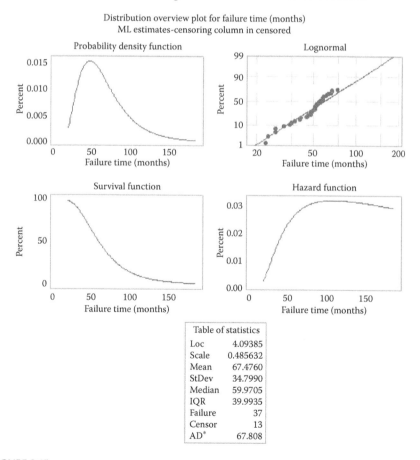

Distribution overview plot for failure time (months)
ML estimates-censoring column in censored

Table of statistics	
Loc	4.09385
Scale	0.485632
Mean	67.4760
StDev	34.7990
Median	59.9705
IQR	39.9935
Failure	37
Censor	13
AD*	67.808

FIGURE 3.15
Step 6 in Section 3.2.

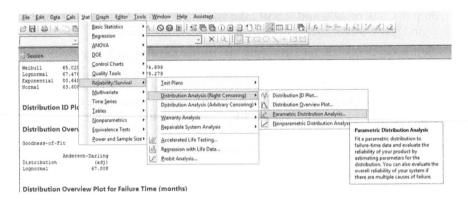

FIGURE 3.16
Step 7 commands in Section 3.2.

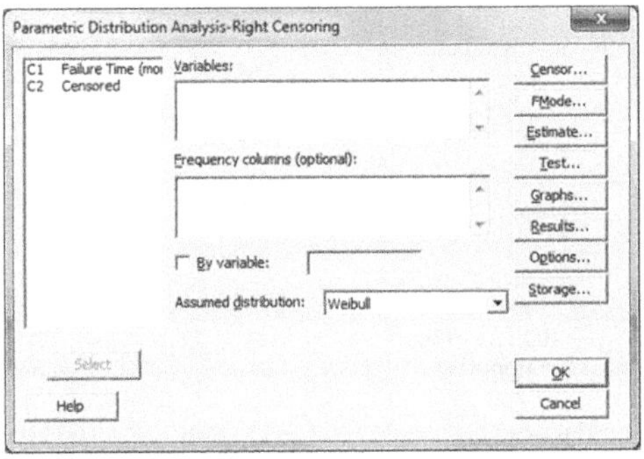

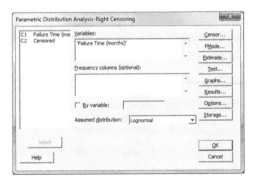

FIGURE 3.17
Step 7 dialog box in Section 3.2.

FIGURE 3.18
Step 8 in Section 3.2.

Step 9: Click on *Censor,* and the dialog box shown in Figure 3.19 opens.

Step 10: As shown in Figure 3.20, select the "Censored" column for *Use censoring columns* and enter "0" for *Censoring value.*

Step 11: Click on *OK,* and it will take you back to the dialog box shown in Figure 3.18.

Step 12: Click on *Estimate,* and the dialog box shown in Figure 3.21 opens.

Step 13: As shown in Figure 3.22, enter "70" for *Estimate probabilities for these times (values).*

Step 14: Click on *OK,* and it will take you back to the dialog box shown in Figure 3.18.

Step 15: Click on *Graphs,* and the dialog box shown in Figure 3.23 opens.

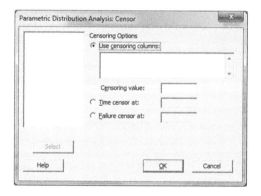

FIGURE 3.19
Step 9 in Section 3.2.

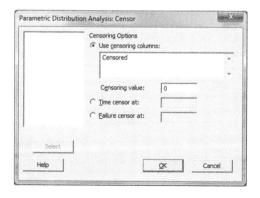

FIGURE 3.20
Step 10 in Section 3.2.

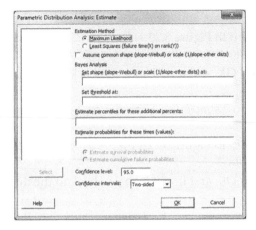

FIGURE 3.21
Step 12 in Section 3.2.

FIGURE 3.22
Step 13 in Section 3.2.

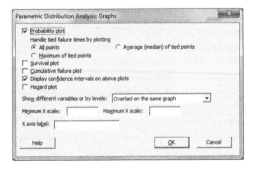

FIGURE 3.23
Step 15 in Section 3.2.

Step 16: As shown in Figure 3.24, uncheck all boxes and check the box for *Survival Plot*. Click on *OK*, and it will take you back to the dialog box shown in Figure 3.18.

Step 17: Click on *OK*, and you will see the graph shown in Figure 3.25.

Step 18: As shown in Figure 3.26, put the cursor anywhere on the survival plot curve to see the survival probabilities at different times. For example, 79.7834% of the fans are expected to survive past 40 months.

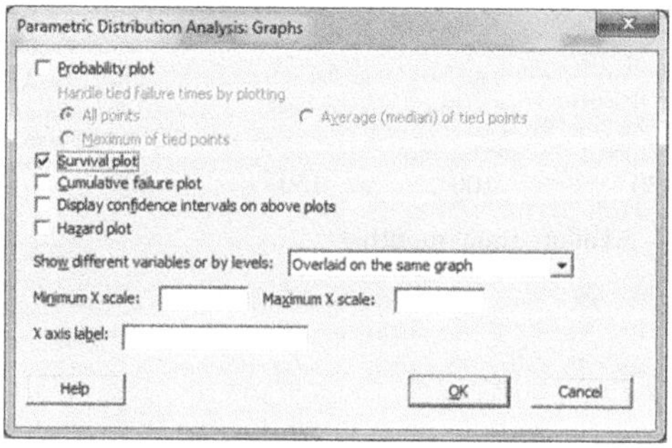

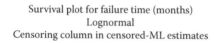

FIGURE 3.24
Step 16 in Section 3.2.

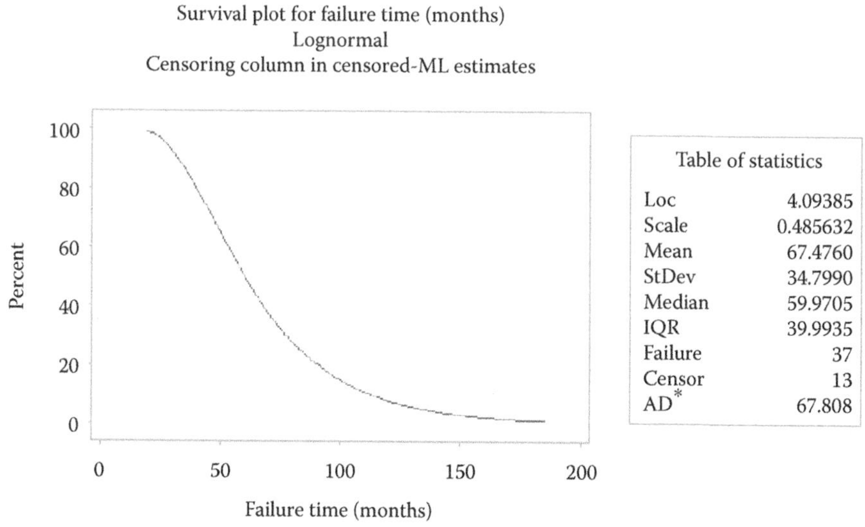

Survival plot for failure time (months)
Lognormal
Censoring column in censored-ML estimates

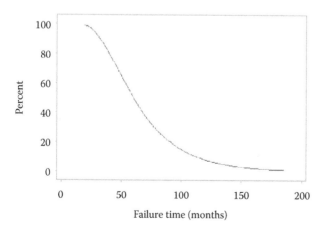

Table of statistics	
Loc	4.09385
Scale	0.485632
Mean	67.4760
StDev	34.7990
Median	59.9705
IQR	39.9935
Failure	37
Censor	13
AD*	67.808

FIGURE 3.25
Step 17 in Section 3.2.

Survival plot for failure time (months)
Lognormal
Censoring column in censored-ML estimates

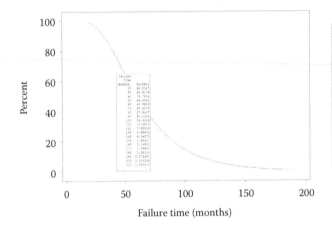

FIGURE 3.26
Step 18 in Section 3.2.

In addition to the survival plot shown in Figure 3.25, you will see the following output in the session window:

```
Table of Survival Probabilities

                             95.0% Normal CI
Time   Probability      Lower      Upper
 70        0.375078   0.263837   0.497882
```

Based on this output, 37.5078% of the fans are expected to survive past 70 months.

3.3 Conclusion

This case study illustrated how to perform reliability analysis of a product with right-censored and exact failure time data. Minitab was used to determine the statistical distribution that best fits the failure time data and then to perform appropriate analysis with that distribution.

4

Reliability Analysis with Arbitrarily Censored Failure Times

Recall from Chapter 1 that arbitrarily censored failure times may consist of

- Right-censored failure time (the unit has not failed by that time, but we know that it will fail at anytime later)
- Left-censored failure time (the unit failed by that time, but we do not know exactly when)
- Exact failure time (the unit failed exactly at that time)
- Interval-censored failure time (the unit failed within a time *interval*, but we do not know exactly when)

In this case study, a reliability engineer at a supplier of diesel engine parts wishes to perform reliability analysis on crankshafts. Specifically, she is interested in knowing the answers to the following questions:

- By what time will 60% of the crankshafts fail?
- By what time will 99% of the crankshafts fail?
- What percentage of crankshafts will survive past 45,000 engine cycles?

Table 4.1 shows the failure time data from a test conducted on 776 randomly selected crankshafts. Notice from the table that 11 crankshafts have failed by 10,000 cycles (left-censored), 71 crankshafts survived past 90,000 cycles (right-censored), and the remaining crankshafts failed in different intervals of time. There are no exact-censored failure times in this table, but if, say, 14 crankshafts failed exactly at the end of, say, 13,000 cycles, we would have entered 13,000 cycles in both Start and End columns, and 14 crankshafts in the Frequency column.

Section 4.1 illustrates the approach to identify the statistical distribution that best fits the failure time data in Table 4.1. Section 4.2 gives the approach to perform reliability analysis of the data. Section 4.3 gives an additional example with a different data set. Finally, Section 4.4 gives some conclusions.

TABLE 4.1

Failure Time Data for Sections 4.1 and 4.2

Start	End	Frequency
*	10,000	11
10,000	20,000	10
20,000	30,000	14
30,000	40,000	25
40,000	50,000	37
50,000	60,000	87
60,000	70,000	145
70,000	80,000	231
80,000	90,000	145
90,000	*	71

4.1 Determination of Statistical Distribution

Perform the following steps to determine the statistical distribution that best fits the failure time data:

Step 1: Open a new Minitab® project file and copy the data from Table 4.1, as shown in Figure 4.1.

Step 2: As shown in Figure 4.2, go to *Stat* → *Reliability/Survival* → *Distribution Analysis (Arbitrary Censoring)* → *Distribution ID Plot*. The dialog box shown in Figure 4.3 opens.

Step 3: As shown in Figure 4.4, select the "Start" column for *Start variables,* select the "End" column for *End variables,* select the "Frequency" column for *Frequency columns,* and select the four distributions: Weibull, Lognormal, Exponential, and Smallest extreme value. (Another option is to select *Use all distributions,* which will make Minitab test a total of 11 distributions.)

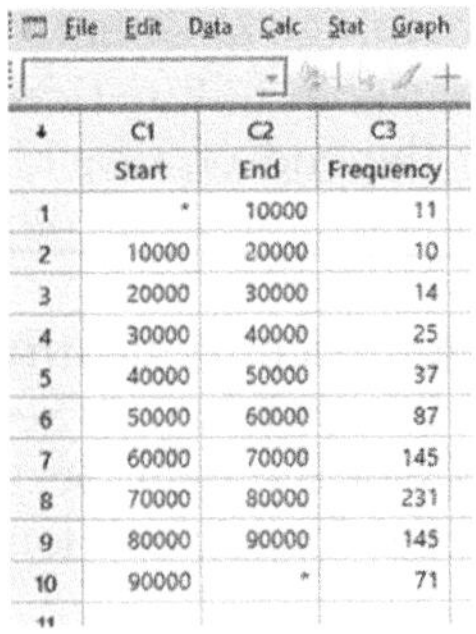

	C1 Start	C2 End	C3 Frequency
1	*	10000	11
2	10000	20000	10
3	20000	30000	14
4	30000	40000	25
5	40000	50000	37
6	50000	60000	87
7	60000	70000	145
8	70000	80000	231
9	80000	90000	145
10	90000	*	71

FIGURE 4.1
Step 1 in Section 4.1.

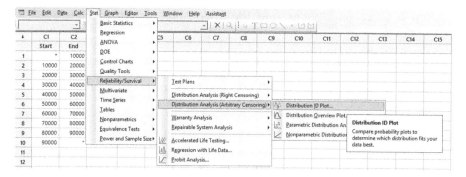

FIGURE 4.2
Step 2 commands in Section 4.1.

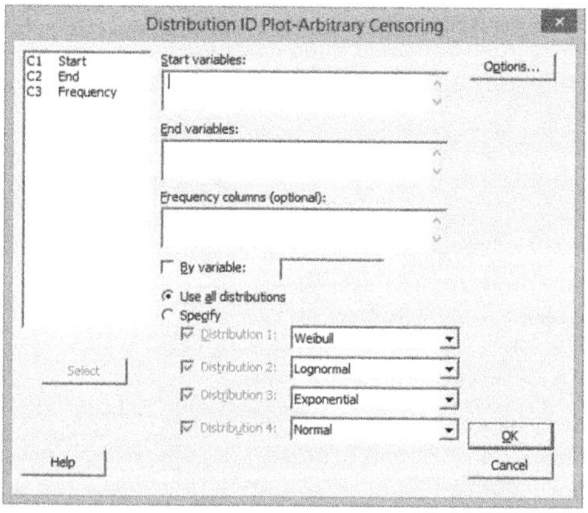

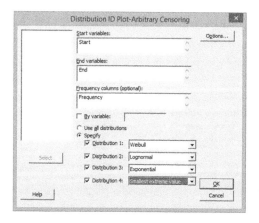

FIGURE 4.3
Step 2 dialog box in Section 4.1.

FIGURE 4.4
Step 3 in Section 4.1.

Step 4: Click on *Options*, and the dialog box shown in Figure 4.5 opens.

Step 5: As shown in Figure 4.6, add "60" and "99" in the box for *Estimate percentiles for these percents*.

Step 6: Click on *OK*, and it will take you back to the dialog box shown in Figure 4.4.

Step 7: Click on *OK*, and you will see the graph shown in Figure 4.7.

Notice from the graph in Figure 4.7 that the smallest extreme value distribution has the lowest *Anderson–Darling (adj)* value (2.408) among all four distributions. This is because the smallest extreme value distribution best fits the failure time data, among all four distributions. Notice that the points are more close to the straight line for the smallest extreme value distribution than they are to the straight lines for the other three distributions.

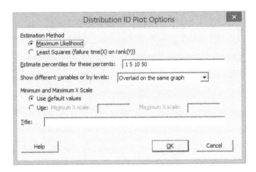

FIGURE 4.5
Step 4 in Section 4.1.

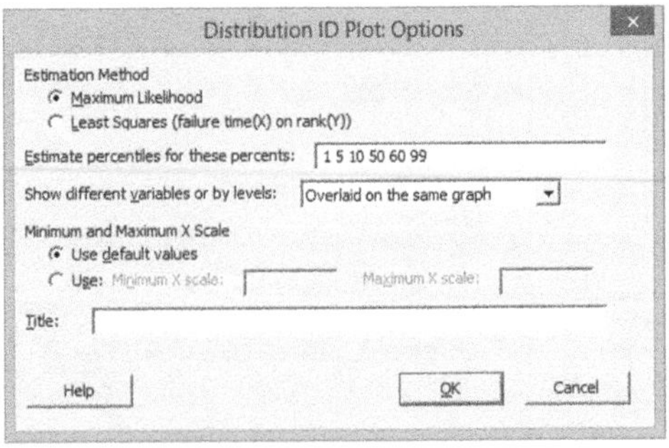

FIGURE 4.6
Step 5 in Section 4.1.

In addition to the graph shown in Figure 4.7, you will see the following output in the session window:

```
Table of Percentiles

                                          Standard    95% Normal CI
Distribution               Percent  Percentile   Error    Lower     Upper
Weibull                          1    26700.1   995.958  24817.7   28725.3
Lognormal                        1    25990.8   777.921  24510.0   27561.1
Exponential                      1    759.367   28.6203  705.294   817.586
Smallest Extreme Value           1    12165.8   2251.58  7752.80   16578.8

Weibull                          5    38680.3   983.807  36799.3   40657.4
Lognormal                        5    34254.2   804.360  32713.4   35867.5
Exponential                      5    3875.54   146.068  3599.57   4172.66
Smallest Extreme Value           5    35289.4   1546.31  32258.7   38320.1

Weibull                         10    45559.7   928.686  43775.3   47416.7
Lognormal                       10    39685.2   810.699  38127.6   41306.3
Exponential                     10    7960.66   300.035  7393.80   8570.98
Smallest Extreme Value          10    45501.4   1246.75  43057.8   47944.9

Weibull                         50    69924.9   671.909  68620.3   71254.3
Lognormal                       50    66694.1   987.627  64786.2   68658.2
Exponential                     50    52371.7   1973.87  48642.4   56386.9
Smallest Extreme Value          50    72226.9   607.153  71036.9   73416.9

Weibull                         60    74506.7   663.124  73218.3   75817.8
Lognormal                       60    73902.3   1122.90  71734.0   76136.3
Exponential                     60    69231.6   2609.32  64301.8   74539.4
Smallest Extreme Value          60    76186.3   562.609  75083.6   77289.0

Weibull                         99     107561   1440.54   104774    110422
Lognormal                       99     171141   5288.13   161084    181826
Exponential                     99     347950   13114.1   323173    374626
Smallest Extreme Value          99    99092.1   800.575  97523.0    100661


Table of MTTF

                                Standard    95% Normal CI
Distribution             Mean     Error    Lower     Upper
Weibull                69265.3   642.37   68017.7   70535.9
Lognormal              72397.0   1135.09  70206.1   74656.3
Exponential            75556.4   2847.70  70176.2   81349.1
Smallest Extreme Value 69237.8   655.05   67953.9   70521.7
```

From this output, it is evident that

- 1% of the crankshafts are expected to fail in 12,165.8 cycles.
- 5% of the crankshafts are expected to fail in 35,289.4 cycles.
- 10% of the crankshafts are expected to fail in 45,501.4 cycles.
- 50% of the crankshafts are expected to fail in 72,226.9 cycles.
- 60% of the crankshafts are expected to fail in 76,186.3 cycles.
- 99% of the crankshafts are expected to fail in 99,092.1 cycles.
- On average, a crankshaft is expected to fail in 69,237.8 cycles (mean time to failure, MTTF).

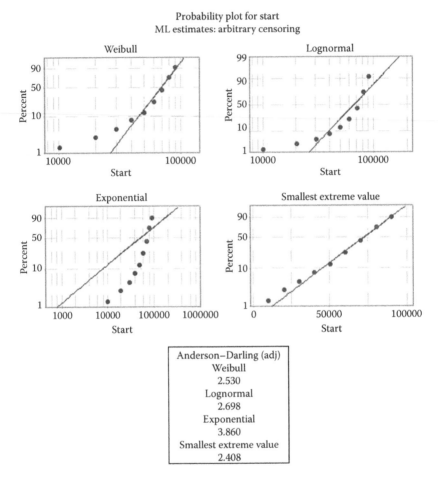

FIGURE 4.7
Step 7 in Section 4.1.

4.2 Reliability Analysis

Perform the following steps to carry out the reliability analysis of the data:

Step 1: As shown in Figure 4.8, go to *Stat → Reliability/Survival → Distribution Analysis (Arbitrary Censoring) → Distribution Overview Plot.* The dialog box shown in Figure 4.9 opens.

Step 2: As shown in Figure 4.10, select the "Start" column for *Start variables,* select the "End" column for *End variables,* select the "Frequency" column for *Frequency columns,* and select "Smallest extreme value" for *Parametric analysis.*

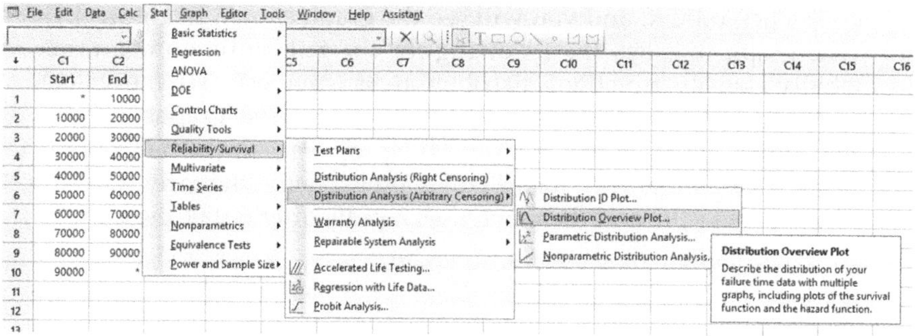

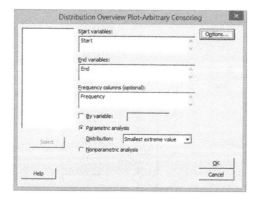

FIGURE 4.8
Step 1 commands in Section 4.2.

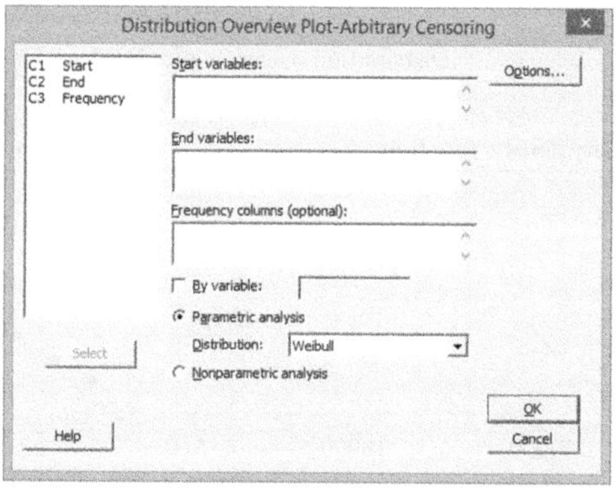

FIGURE 4.9
Step 1 dialog box in Section 4.2.

FIGURE 4.10
Step 2 in Section 4.2.

Step 3: Click on *OK*, and you will see the graph shown in Figure 4.11. The graph displays the following for the failure time data: probability density function, distribution fit (smallest extreme value), survival function, and hazard rate function. These graphs help the reliability engineer get a sense of the properties of the failure time data. For example, the hazard rate is almost flat in the early stages of the product's life and then increases rapidly after 60,000 cycles.

Step 4: As shown in Figure 4.12, go to *Stat → Reliability/Survival → Distribution Analysis (Arbitrary Censoring) → Parametric Distribution Analysis*. The dialog box shown in Figure 4.13 opens.

Step 5: As shown in Figure 4.14, select the "Start" column for *Start variables*, select the "End" column for *End variables*, select the "Frequency" column for *Frequency columns*, and select "Smallest extreme value" for *Assumed distribution*.

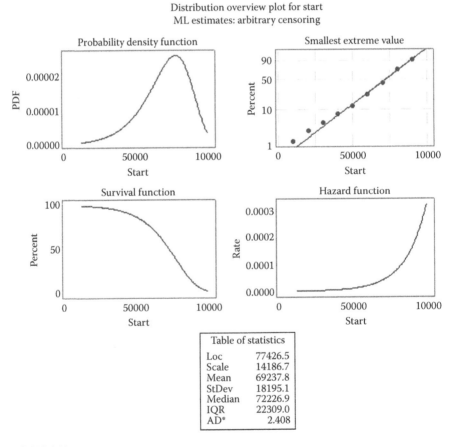

Distribution overview plot for start
ML estimates: arbitrary censoring

Table of statistics	
Loc	77426.5
Scale	14186.7
Mean	69237.8
StDev	18195.1
Median	72226.9
IQR	22309.0
AD*	2.408

FIGURE 4.11
Step 3 in Section 4.2.

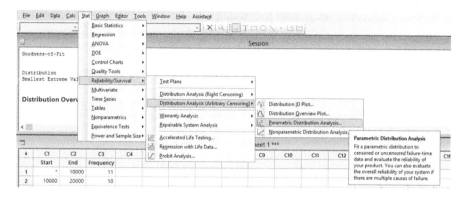

FIGURE 4.12
Step 4 commands in Section 4.2.

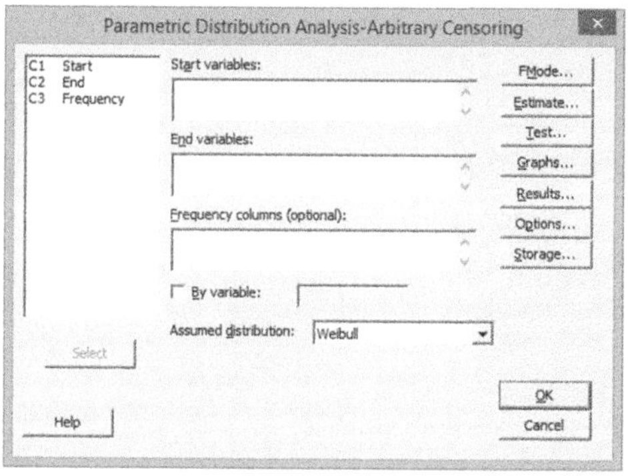

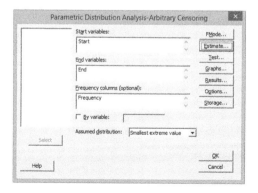

FIGURE 4.13
Step 4 dialog box in Section 4.2.

FIGURE 4.14
Step 5 in Section 4.2.

Step 6: Click on *Estimate*, and the dialog box shown in Figure 4.15 opens.

Step 7: As shown in Figure 4.16, enter "45000" for *Estimate probabilities for these times (values)*.

Step 8: Click on *OK*, and it will take you back to the dialog box shown in Figure 4.14.

Step 9: Click on *Graphs*, and the dialog box shown in Figure 4.17 opens.

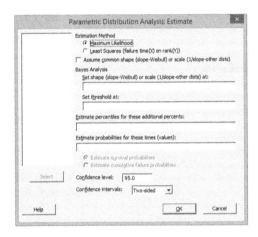

FIGURE 4.15
Step 6 in Section 4.2.

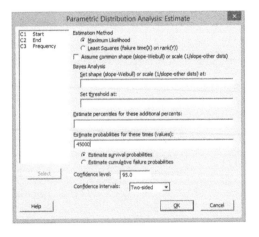

FIGURE 4.16
Step 7 in Section 4.2.

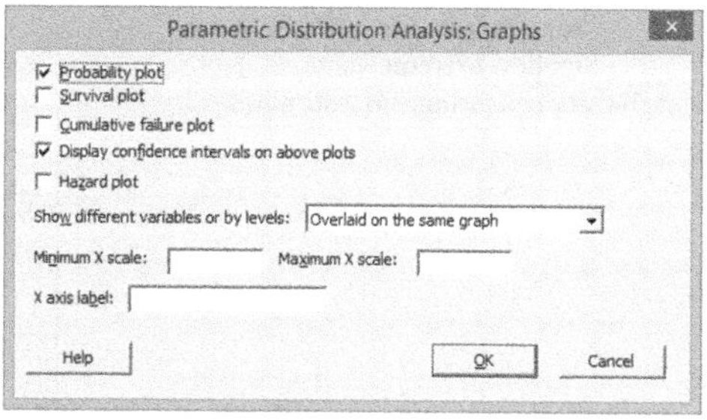

FIGURE 4.17
Step 9 in Section 4.2.

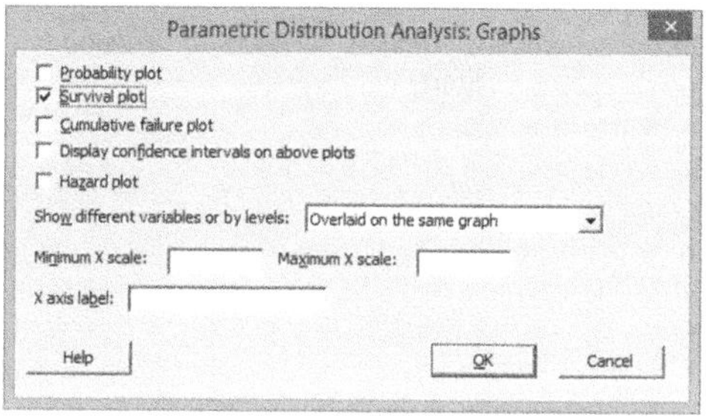

FIGURE 4.18
Step 10 in Section 4.2.

Step 10: As shown in Figure 4.18, uncheck all boxes, and check the box for *Survival plot*. Click on *OK*, and it will take you back to the dialog box shown in Figure 4.14.

Step 11: Click on *OK*, and you will see the graph shown in Figure 4.19.

Step 12: As shown in Figure 4.20, put the cursor anywhere on the survival plot curve to see the survival probabilities at different times. For example, 58.6838% of the crankshafts are expected to survive past 68,500 cycles.

In addition to the survival plot shown in Figure 4.19, you will see the following output in the session window:

```
Table of Survival Probabilities

                            95.0% Normal CI
  Time    Probability     Lower       Upper
 45000       0.903299   0.885981   0.918108
```

Based on this output, 90.3299% of the crankshafts are expected to survive past 45,000 cycles.

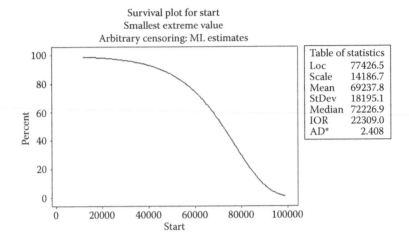

FIGURE 4.19
Step 11 in Section 4.2.

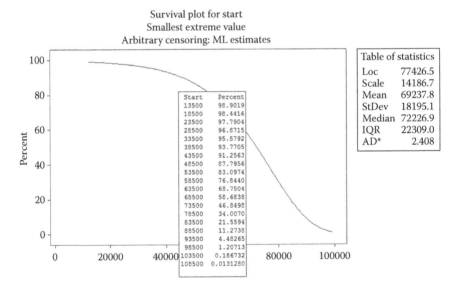

FIGURE 4.20
Step 12 in Section 4.2.

4.3 Additional Example

In this example, a reliability engineer at a supplier of gasoline engine parts wishes to perform reliability analysis on crankshafts. Unlike Section 4.2, exact failures are also considered here. Specifically, the engineer is interested in knowing the answers to the following questions:

- By what time will 70% of the crankshafts fail?
- By what time will 99.9% of the crankshafts fail?
- What percentage of crankshafts will survive past 42,000 engine cycles?

Table 4.2 shows the failure time data from a test conducted on 800 randomly selected crankshafts. Notice from the table that 11 crankshafts have failed by 10,000 cycles (left-censored), 71 crankshafts survived past 90,000 cycles (right-censored), 24 crankshafts failed at the end of exactly 13,000 cycles, and the remaining crankshafts failed in different intervals of time.

Perform the following steps to determine the statistical distribution that best fits the failure time data:

Step 1: Open a new Minitab project file and copy the data from Table 4.2, as shown in Figure 4.21.

Step 2: As shown in Figure 4.22, go to *Stat → Reliability/Survival → Distribution Analysis (Arbitrary Censoring) → Distribution ID Plot*. The dialog box shown in Figure 4.23 opens.

Step 3: As shown in Figure 4.24, select the "Start" column for *Start variables*, select the "End" column for *End variables*, select the "Frequency" column for *Frequency columns*, and select *Use all distributions*.

TABLE 4.2

Failure Time Data for Section 4.3

Start	End	Frequency
*	10,000	11
10,000	20,000	10
13,000	13,000	24
20,000	30,000	14
30,000	40,000	25
40,000	50,000	37
50,000	60,000	87
60,000	70,000	145
70,000	80,000	231
80,000	90,000	145
90,000	*	71

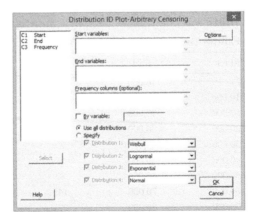

FIGURE 4.21
Step 1 in Section 4.3.

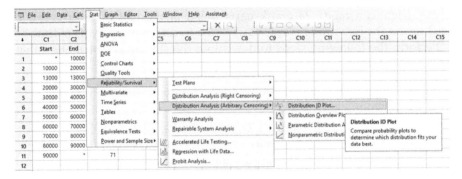

FIGURE 4.22
Step 2 commands in Section 4.3.

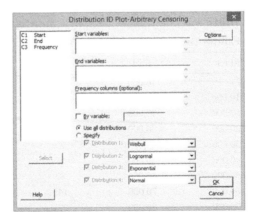

FIGURE 4.23
Step 2 dialog box in Section 4.3.

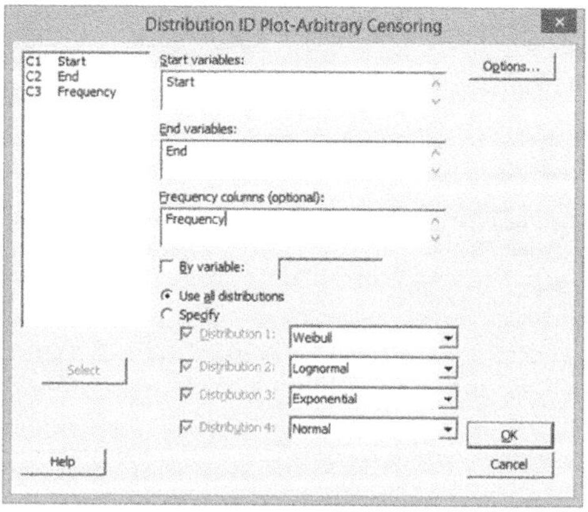

FIGURE 4.24
Step 3 in Section 4.3.

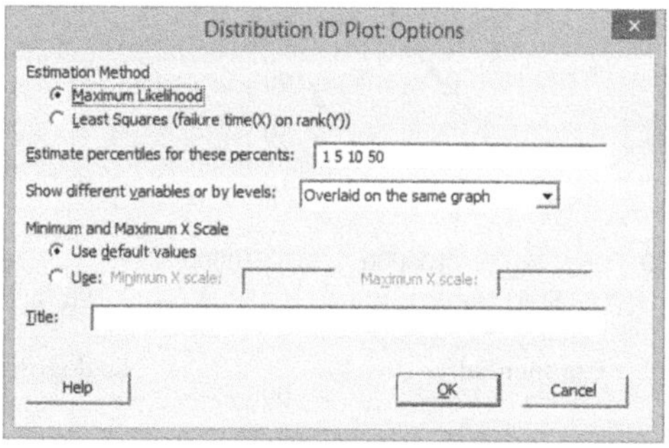

FIGURE 4.25
Step 4 in Section 4.3.

Step 4: Click on *Options*, and the dialog box shown in Figure 4.25 opens.

Step 5: As shown in Figure 4.26, add "70" and "99.9" in the box for *Estimate percentiles for these percents*.

Step 6: Click on *OK*, and it will take you back to the dialog box shown in Figure 4.24.

Step 7: Click on *OK*, and you will see the three graphs shown in Figures 4.27 through 4.29.

Notice from the graphs in Figures 4.27 through 4.29 that three-parameter loglogistic distribution has the lowest *Anderson–Darling (adj)* value (2.207) among all 11 distributions. This is because the three-parameter loglogistic distribution best fits the failure time data, among all of the 11 distributions. Notice that the points are more close to the straight line for the three-parameter loglogistic distribution than they are to the straight lines for the other 10 distributions.

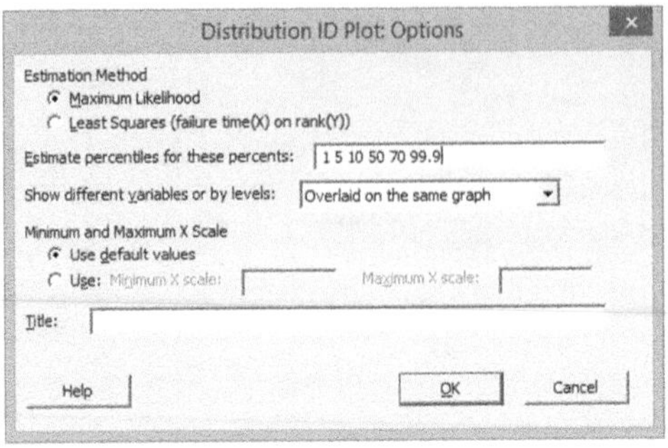

FIGURE 4.26
Step 5 in Section 4.3.

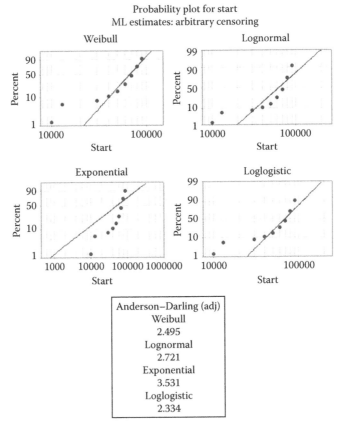

FIGURE 4.27
Step 7 Graph Set 1 in Section 4.3.

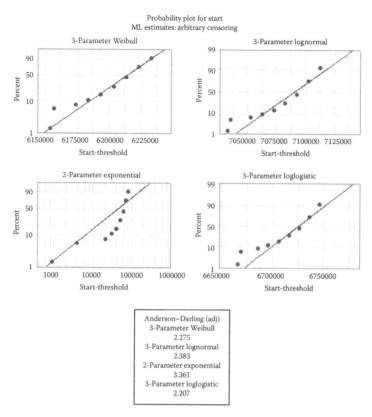

FIGURE 4.28
Step 7 Graph Set 2 in Section 4.3.

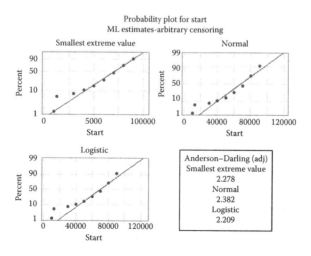

FIGURE 4.29
Step 7 Graph Set 3 in Section 4.3.

In addition to the graphs shown in Figure 4.27 through 4.29, you will see the following output in the session window:

Weibull	70	78656.5	790.011	77123.2	80220.2
Lognormal	70	82847.1	1628.02	79716.9	86100.2
Exponential	70	88484.7	3279.66	82284.6	95152.0
Loglogistic	70	82549.4	1151.28	80323.5	84837.0
3-Parameter Weibull	70	79511.8	582.641	78369.9	80653.8
3-Parameter Lognormal	70	79407.1	853.529	77734.2	81080.0
2-Parameter Exponential	70	85619.5	2837.34	80235.2	91365.2
3-Parameter Loglogistic	70	79816.6	755.527	78335.8	81297.4
Smallest Extreme Value	70	79502.4	581.135	78363.4	80641.4
Normal	70	79396.3	850.481	77729.4	81063.2
Logistic	70	79786.9	753.234	78310.5	81263.2
Weibull	99.9	125802	2358.43	121264	130511
Lognormal	99.9	299637	13982.4	273448	328334
Exponential	99.9	507678	18816.9	472105	545931
Loglogistic	99.9	314591	16369.4	284089	348367
3-Parameter Weibull	99.9	106597	1023.76	104591	108604
3-Parameter Lognormal	99.9	135296	2038.99	131300	139292
2-Parameter Exponential	99.9	448032	16365.7	417077	481285
3-Parameter Loglogistic	99.9	150099	2662.50	144881	155318
Smallest Extreme Value	99.9	106471	1015.44	104481	108461
Normal	99.9	134875	2015.33	130925	138825
Logistic	99.9	149508	2625.91	144362	154655

Table of MTTF

Distribution	Mean	Standard Error	95% Normal CI Lower	Upper
Weibull	67548.8	714.35	66163.1	68963.5
Lognormal	72224.0	1410.13	69512.4	75041.3
Exponential	73493.9	2724.03	68344.3	79031.7
Loglogistic	74279.6	1016.91	72313.0	76299.7
3-Parameter Weibull	67753.3	699.10	66383.1	69123.5
3-Parameter Lognormal	68071.5	780.05	66542.6	69600.3
2-Parameter Exponential	72659.3	2355.74	68185.8	77426.3
3-Parameter Loglogistic	70079.4	702.32	68702.9	71455.9
Smallest Extreme Value	67726.4	700.42	66353.6	69099.2
Normal	68057.6	778.52	66531.7	69583.5
Logistic	70037.7	701.70	68662.3	71413.0

From this output, it is evident that

- 70% of the crankshafts are expected to fail in 79,816.6 cycles.
- 99.9% of the crankshafts are expected to fail in 150,099 cycles.
- On average, a crankshaft is expected to fail in 70,079.4 cycles (mean time to failure, MTTF).

Perform the following steps to carry out the reliability analysis of the data:

Step 1: As shown in Figure 4.30, go to *Stat → Reliability/Survival → Distribution Analysis (Arbitrary Censoring) → Distribution Overview Plot*. The dialog box shown in Figure 4.31 opens.

Step 2: As shown in Figure 4.32, select the "Start" column for *Start variables*, select the "End" column for *End variables*, select the "Frequency"

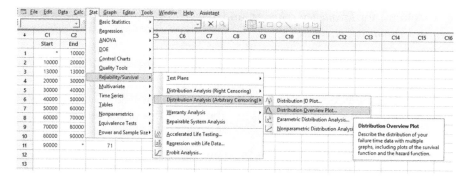

FIGURE 4.30
Step 1 commands for distribution overview plot in Section 4.3.

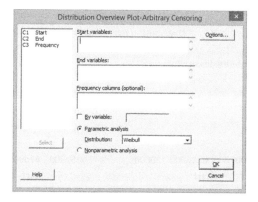

FIGURE 4.31
Step 1 dialog box for distribution overview plot in Section 4.3.

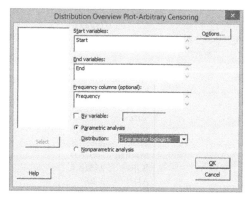

FIGURE 4.32
Step 2 for distribution overview plot in Section 4.3.

column for *Frequency columns,* and select "3-parameter loglogistic" for *Parametric analysis.*

Step 3: Click on *OK,* and you will see the graph shown in Figure 4.33. The graph displays the following for the failure time data: probability density function, distribution fit (three-parameter loglogistic), survival function, and hazard rate function. These graphs help the reliability engineer get a sense of the properties of the failure time data.

Step 4: As shown in Figure 4.34, go to *Stat → Reliability/Survival → Distribution Analysis (Arbitrary Censoring) → Parametric Distribution Analysis.* The dialog box shown in Figure 4.35 opens.

Step 5: As shown in Figure 4.36, select the "Start" column for *Start variables,* select the "End" column for *End variables,* select the "Frequency"

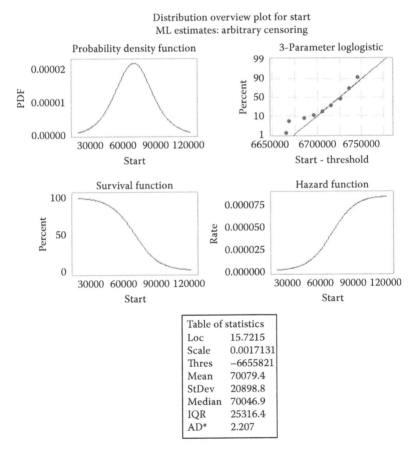

FIGURE 4.33
Step 3 for distribution overview plot in Section 4.3.

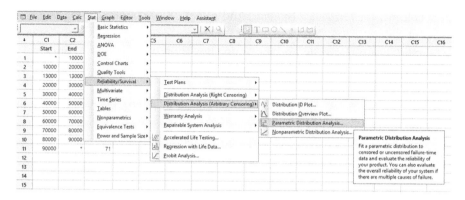

FIGURE 4.34
Step 4 commands for parametric analysis in Section 4.3.

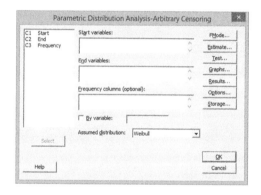

FIGURE 4.35
Step 4 dialog box for parametric analysis in Section 4.3.

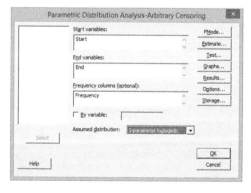

FIGURE 4.36
Step 5 for parametric analysis in Section 4.3.

column for *Frequency columns,* and select "3-parameter loglogistic" for *Assumed distribution.*

Step 6: Click on *Estimate,* and the dialog box shown in Figure 4.37 opens.

Step 7: As shown in Figure 4.38, enter "42000" for *Estimate probabilities for these times (values).*

Step 8: Click on *OK,* and it will take you back to the dialog box shown in Figure 4.36.

Step 9: Click on *Graphs,* and the dialog box shown in Figure 4.39 opens.

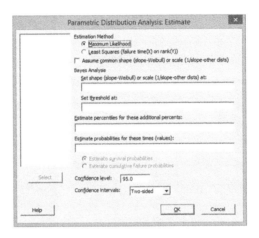

FIGURE 4.37
Step 6 for parametric analysis in Section 4.3.

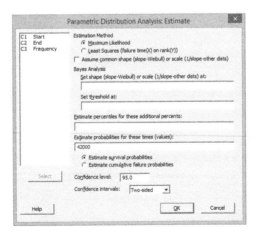

FIGURE 4.38
Step 7 for parametric analysis in Section 4.3.

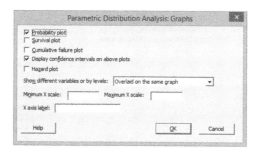

FIGURE 4.39
Step 9 for parametric analysis in Section 4.3.

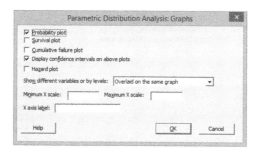

FIGURE 4.40
Step 10 for parametric analysis in Section 4.3.

Step 10: As shown in Figure 4.40, uncheck all boxes and check the box for *Survival Plot*. Click on *OK* and it will take you back to the dialog box shown in Figure 4.36.

Step 11: Click on *OK*, and you will see the graph shown in Figure 4.41.

Step 12: As shown in Figure 4.42, put the cursor anywhere on the survival plot curve to see the survival probabilities at different times. For example, 95.0822% of the crankshafts are expected to survive past 36,000 cycles.

In addition to the survival plot shown in Figure 4.42, you will see the following output in the session window:

```
Table of Survival Probabilities

                       95.0% Normal CI
   Time   Probability     Lower      Upper
  42000      0.919733   0.903633   0.933341
```

Based on this output, 91.9733% of the crankshafts are expected survive past 42,000 cycles.

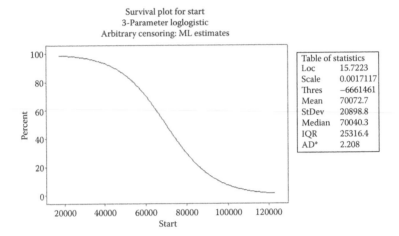

FIGURE 4.41
Step 11 for parametric analysis in Section 4.3.

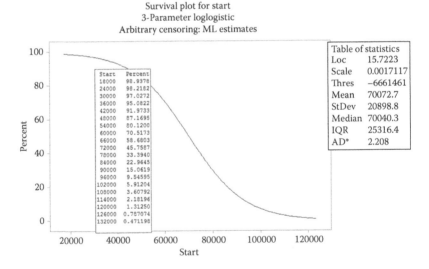

FIGURE 4.42
Step 12 for parametric analysis in Section 4.3.

4.4 Conclusion

These case studies illustrated how to perform reliability analysis of units with arbitrarily censored failure time data. Minitab was used to determine the statistical distributions that best fit the failure time data and then to perform appropriate analysis with those distributions.

5

Nonparametric Reliability Analysis

Recall from Chapters 3 and 4 that the statistical distribution with the lowest *Anderson–Darling (adj)* value among all the considered distributions is the best fit for the failure time data. However, it is possible that none of the considered distributions is a good fit because the points are not close to the straight line in the probability distribution plot of any of the distributions. In such a case, nonparametric reliability analysis may be performed.

In this case study, a reliability engineer at a firm that manufactures mechanical motors wishes to perform reliability analysis on the products. First, we will perform the analysis with exact and right-censored failure time data. Then, we will perform the analysis with arbitrarily censored failure time data.

Recall from Chapter 1 that a right-censored failure time means that the product that is being tested has not failed by that time. For example, if a product has not failed in, say, 100 hours, and if the test duration has ended, the failure time is recorded as 100 hours (right-censored). This product has not failed in 100 hours, but is expected to fail anytime later. Recall also that an exact failure time means that the product has failed exactly at that time. For example, if a product has failed exactly at the end of, say, 50 hours, the failure time is recorded as 50 hours (exact).

Similarly, arbitrarily censored failure times may consist of

- Right-censored failure time (the product has not failed by that time, but we know that it will fail at anytime later)
- Left-censored failure time (the product failed by that time, but we do not know exactly when)
- Exact failure time (the product failed exactly at that time)
- Interval-censored failure time (the product failed within a time *interval*, but we do not know exactly when)

Section 5.1 illustrates the nonparametric analysis with exact and right-censored failure time data. Section 5.2 gives the nonparametric analysis with arbitrarily censored failure time data. Finally, Section 5.3 gives some conclusions.

5.1 Nonparametric Analysis with Exact and Right-Censored Failure Times

Table 5.1 shows the failure time data from a test conducted on 40 randomly selected motors. In the "Censored" column of this table, 1 refers to exact failure and 0 refers to right-censored.

We shall assume here that performing the steps listed in Section 3.1 have resulted in probability plots where none of the considered distributions is a good fit, that is, the points are not close to the straight line in the probability distribution plot of any of the distributions.

The reliability engineer is interested in knowing the answers to the following questions:

- By what time are 50% of the motors expected to fail?
- What proportion of the motors are expected to survive past 14 months?
- By what time are 20% of the motors expected to fail?

Perform the following steps for nonparametric reliability analysis:

Step 1: Open a new Minitab® project file and copy the data from Table 5.1, as shown in Figure 5.1.

Step 2: As shown in Figure 5.2, go to *Stat → Reliability/Survival → Distribution Analysis (Right Censoring) → Nonparametric Distribution Analysis*. The dialog box shown in Figure 5.3 opens.

Step 3: As shown in Figure 5.4, select the "Failure Time (months)" column for *Variables*.

Step 4: Click on *Censor,* and the dialog box shown in Figure 5.5 opens.

Step 5: As shown in Figure 5.6, select the "Censored" column for *Use censoring columns* and enter "0" in *Censoring value.*

Step 6: Click on *OK,* and it will take you back to the dialog box shown in Figure 5.4.

Step 7: Click on *Graphs,* and the dialog box shown in Figure 5.7 opens.

Step 8: Check the box for *Survival plot,* as shown in Figure 5.8.

Step 9: Click on *OK,* and it will take you back to the dialog box shown in Figure 5.4.

Step 10: Click on *OK,* and you will see the graph shown in Figure 5.9.

The plot shown in Figure 5.9 can be understood better by looking at the Survival Probability column in the following output that you will see in the session window:

TABLE 5.1

Failure Time Data for Section 5.1

No.	Failure Time (months)	Censored
1	100	0
2	11	1
3	48	1
4	32	1
5	36	1
6	22	1
7	72	1
8	69	1
9	35	1
10	29	1
11	18	1
12	38	1
13	39	1
14	68	1
15	36	1
16	18	1
17	25	1
18	14	1
19	77	0
20	47	1
21	27	1
22	62	0
23	54	1
24	84	0
25	45	1
26	10	1
27	97	0
28	6	1
29	37	1
30	38	1
31	40	1
32	30	1
33	64	0
34	46	1
35	46	1
36	24	1
37	76	1
38	18	1
39	16	1
40	45	1

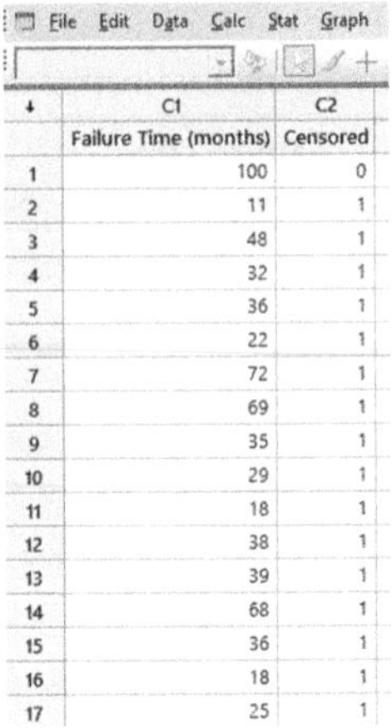

◆	C1	C2
	Failure Time (months)	Censored
1	100	0
2	11	1
3	48	1
4	32	1
5	36	1
6	22	1
7	72	1
8	69	1
9	35	1
10	29	1
11	18	1
12	38	1
13	39	1
14	68	1
15	36	1
16	18	1
17	25	1

FIGURE 5.1
Step 1 in Section 5.1.

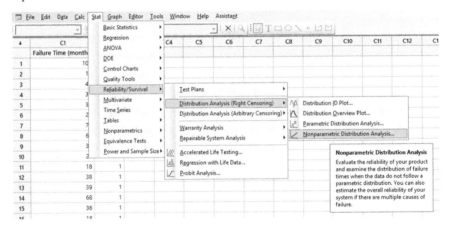

FIGURE 5.2
Step 2 commands in Section 5.1.

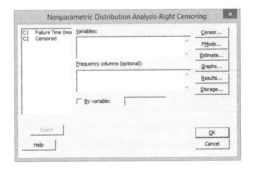

FIGURE 5.3
Step 2 dialog box in Section 5.1.

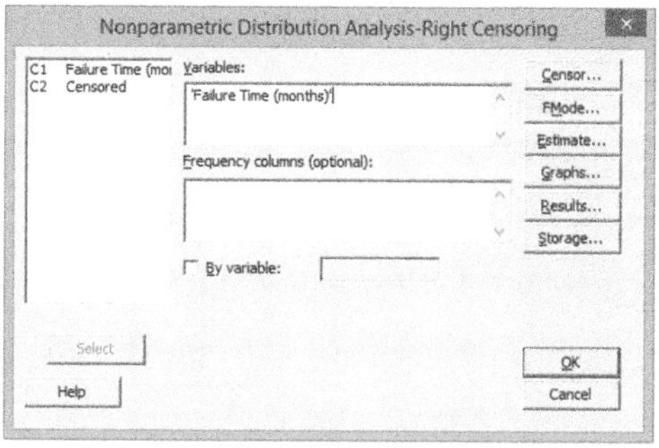

FIGURE 5.4
Step 3 in Section 5.1.

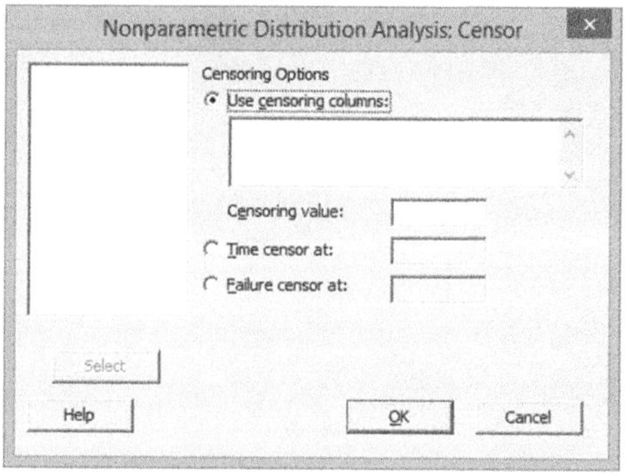

FIGURE 5.5
Step 4 in Section 5.1.

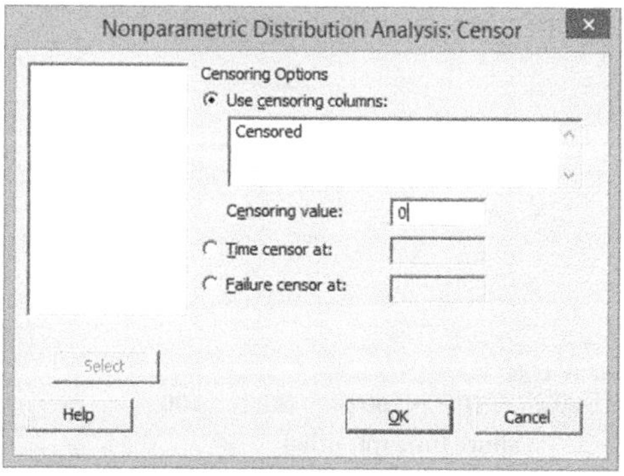

FIGURE 5.6
Step 5 in Section 5.1.

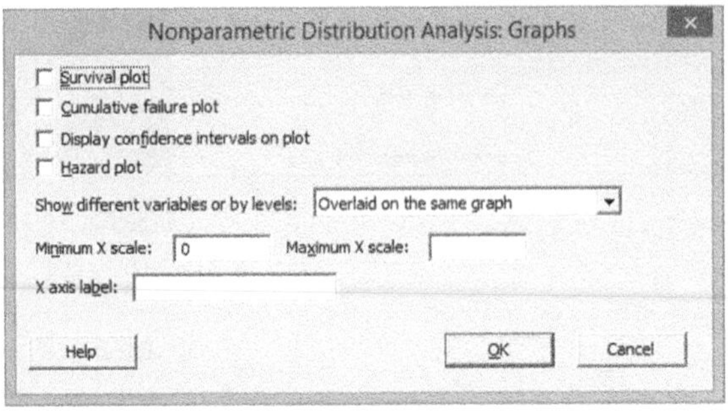

FIGURE 5.7
Step 7 in Section 5.1.

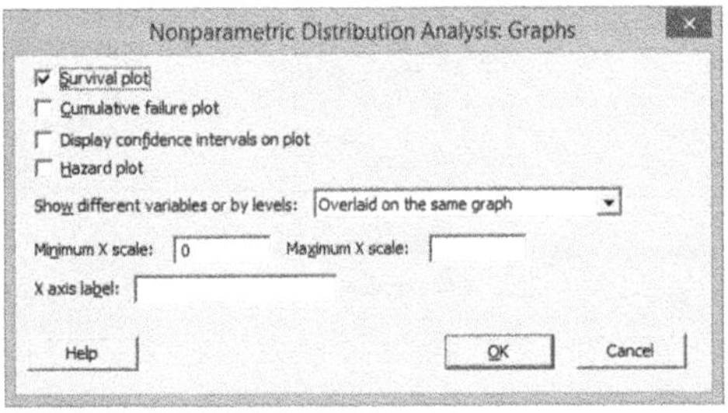

FIGURE 5.8
Step 8 in Section 5.1.

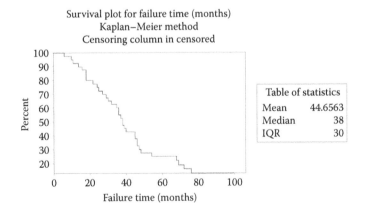

Survival plot for failure time (months)
Kaplan–Meier method
Censoring column in censored

Table of statistics	
Mean	44.6563
Median	38
IQR	30

FIGURE 5.9
Step 10 in Section 5.1.

```
Variable: Failure Time (months)

Censoring Information  Count
Uncensored value          34
Right censored value       6

Censoring value: Censored = 0


Nonparametric Estimates


Characteristics of Variable

           Standard    95.0% Normal CI
Mean(MTTF)   Error   Lower    Upper
   44.6563  4.39032  36.0514  53.2611

Median = 38
IQR = 30   Q1 = 24   Q3 = 54


Kaplan-Meier Estimates

       Number  Number   Survival    Standard    95.0% Normal CI
Time  at Risk  Failed  Probability    Error     Lower     Upper
   6      40      1      0.97500    0.0246855  0.926617  1.00000
  10      39      1      0.95000    0.0344601  0.882459  1.00000
  11      38      1      0.92500    0.0416458  0.843376  1.00000
  14      37      1      0.90000    0.0474342  0.807031  0.99297
  16      36      1      0.87500    0.0522913  0.772511  0.97749
  18      35      3      0.80000    0.0632456  0.676041  0.92396
  22      32      1      0.77500    0.0660256  0.645592  0.90441
  24      31      1      0.75000    0.0684653  0.615810  0.88419
  25      30      1      0.72500    0.0706001  0.586626  0.86337
  27      29      1      0.70000    0.0724569  0.557987  0.84201
  29      28      1      0.67500    0.0740566  0.529852  0.82015
  30      27      1      0.65000    0.0754155  0.502188  0.79781
  32      26      1      0.62500    0.0765466  0.474972  0.77503
  35      25      1      0.60000    0.0774597  0.448182  0.75182
  36      24      2      0.55000    0.0786607  0.395828  0.70417
  37      22      1      0.52500    0.0789581  0.370245  0.67975
  38      21      2      0.47500    0.0789581  0.320245  0.62975
  39      19      1      0.45000    0.0786607  0.295828  0.60417
  40      18      1      0.42500    0.0781625  0.271804  0.57820
  45      17      2      0.37500    0.0765466  0.224972  0.52503
  46      15      2      0.32500    0.0740566  0.179852  0.47015
  47      13      1      0.30000    0.0724569  0.157987  0.44201
  48      12      1      0.27500    0.0706001  0.136626  0.41337
  54      11      1      0.25000    0.0684653  0.115810  0.38419
  68       8      1      0.21875    0.0666585  0.088102  0.34940
  69       7      1      0.18750    0.0640434  0.061977  0.31302
  72       6      1      0.15625    0.0605154  0.037642  0.27486
  76       5      1      0.12500    0.0559017  0.015435  0.23457
```

From this output, it is evident that

- The time by which 50% of the motors are expected to fail is 38 months.
- 90% of the motors are expected to survive past 14 months.
- 20% of the motors are expected to fail in 18 months. (This is because 80% of the motors are expected to survive past 18 months.)

5.2 Nonparametric Analysis with Arbitrarily Censored Failure Times

Table 5.2 shows the failure time data from a test conducted on 776 randomly selected motors. Notice from the table that 11 motors have failed in 10,000 hours (left-censored), 71 motors survived past 90,000 hours (right-censored), and the remaining motors failed in different intervals of time. There are no exact-censored failure times in this table, but if, say, 14 motors failed exactly at the end of, say, 13,000 hours, we would have entered 13,000 hours in both Start and End columns, and 14 motors in the Frequency column.

We shall assume here that performing the steps listed in Section 4.1 have resulted in probability plots where none of the considered distributions is a good fit, that is, the points are not close to the straight line in the probability distribution plot of any of the distributions.

The reliability engineer is interested in knowing the answers to the following questions:

- What percentage of the motors is expected to fail between 60,000 hours and 70,000 hours?
- What percentage of the motors will survive past 40,000 hours?

TABLE 5.2

Failure Time Data for Section 5.2

Start	End	Frequency
*	10,000	11
10,000	20,000	10
20,000	30,000	14
30,000	40,000	25
40,000	50,000	37
50,000	60,000	87
60,000	70,000	145
70,000	80,000	231
80,000	90,000	145
90,000	*	71

Perform the following steps for nonparametric reliability analysis:

Step 1: Open a new Minitab project file and copy the data from Table 5.2, as shown in Figure 5.10.

Step 2: As shown in Figure 5.11, go to *Stat → Reliability/Survival → Distribution Analysis (Arbitrary Censoring) → Nonparametric Distribution Analysis*. The dialog box shown in Figure 5.12 opens.

Step 3: As shown in Figure 5.13, select the "Start" column for *Start variables*, select the "End" column for *End variables*, and select the "Frequency" column for *Frequency columns (optional)*.

	C1 Start	C2 End	C3 Frequency
1	*	10000	11
2	10000	20000	10
3	20000	30000	14
4	30000	40000	25
5	40000	50000	37
6	50000	60000	87
7	60000	70000	145
8	70000	80000	231
9	80000	90000	145
10	90000	*	71

FIGURE 5.10
Step 1 in Section 5.2.

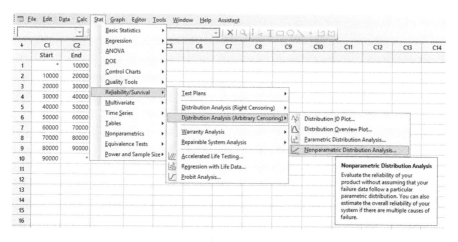

FIGURE 5.11
Step 2 commands in Section 5.2.

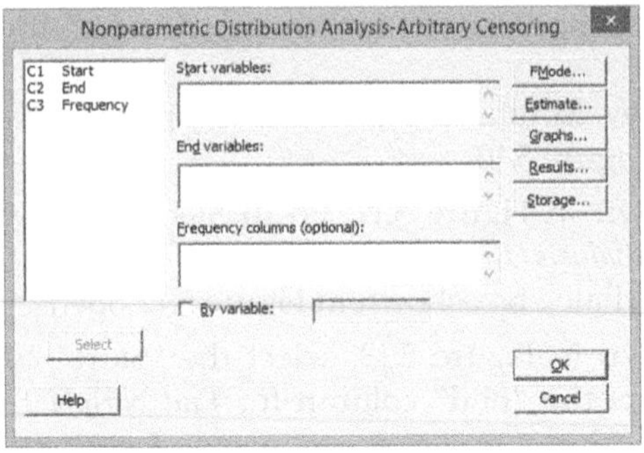

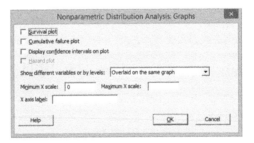

FIGURE 5.12
Step 2 dialog box in Section 5.2.

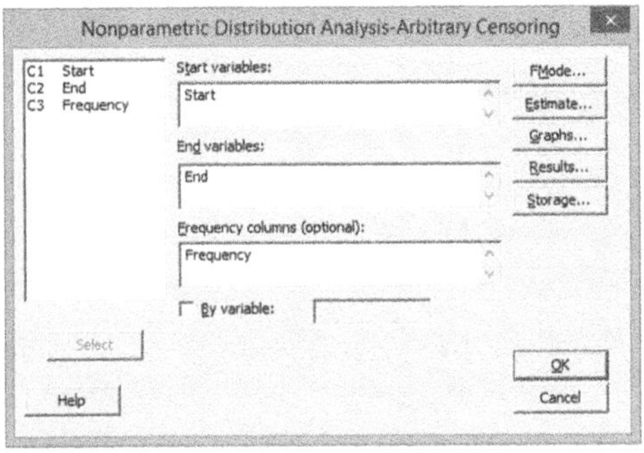

FIGURE 5.13
Step 3 in Section 5.2.

FIGURE 5.14
Step 4 in Section 5.2.

Step 4: Click on *Graphs*, and the dialog box shown in Figure 5.14 opens.

Step 5: Check the box for *Survival plot*, as shown in Figure 5.15.

Step 6: Click on *OK*, and it will take you back to the dialog box shown in Figure 5.13.

Step 7: Click on *OK*, and you will see the graph shown in Figure 5.16.

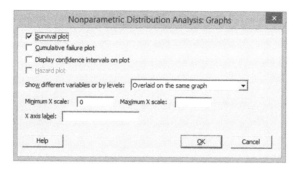

FIGURE 5.15
Step 5 in Section 5.2.

The plot shown in Figure 5.16 can be understood better by looking at the Survival Probability column in the following output that you will see in the session window:

```
Variable Start: Start   End: End
Frequency: Frequency

Censoring Information    Count
Right censored value       71
Interval censored value   694
Left censored value        11


Turnbull Estimates

  Interval      Probability   Standard
Lower   Upper   of Failure     Error
    *   10000    0.014175    0.0042436
10000   20000    0.012887    0.0040488
20000   30000    0.018041    0.0047780
30000   40000    0.032216    0.0063387
40000   50000    0.047680    0.0076495
50000   60000    0.112113    0.0113260
60000   70000    0.186856    0.0139928
70000   80000    0.297680    0.0164139
80000   90000    0.186856    0.0139928
90000     *      0.091495        *


            Survival    Standard     95.0% Normal CI
  Time    Probability    Error      Lower     Upper
10000      0.985825    0.0042436  0.977507  0.994142
20000      0.972938    0.0058249  0.961521  0.984355
30000      0.954897    0.0074499  0.940295  0.969498
40000      0.922680    0.0095883  0.903888  0.941473
50000      0.875000    0.0118721  0.851731  0.898269
60000      0.762887    0.0152678  0.732962  0.792811
70000      0.576031    0.0177402  0.541261  0.610801
80000      0.278351    0.0160890  0.246817  0.309884
90000      0.091495    0.0103498  0.071210  0.111780
```

From this output, it is evident that

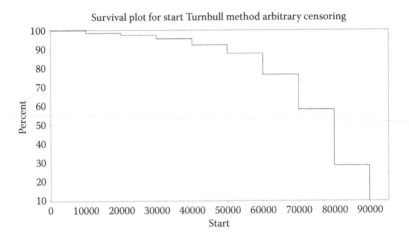

FIGURE 5.16
Step 7 in Section 5.2.

- 18.6856% of the motors are expected to fail between 60,000 hours and 70,000 hours.
- 92.2680% of the motors are expected to survive past 40,000 hours.

5.3 Conclusion

This case study illustrated how to perform nonparametric reliability analysis of a unit. Exact failure times, right-censored failure times, and arbitrarily censored failure times were considered.

6

Warranty Analysis

Many companies may be interested in predicting the amount of money that is needed to cover the warranty costs for their products in a specific period of time in the future. The *Warranty Analysis* option in Minitab® helps in such a situation.

In this case study, a reliability engineer at a firm that manufactures laptop computers wishes to perform warranty analysis on the products. To this end, data regarding sales and warranty claims for the past 12 months have been collected, as shown in Table 6.1. The "Sales" column gives sales volume of 1000 products each month. The columns labeled as "Month 1" through to "Month 12" show the warranty claims processed in each month. For example, the "Month 5" column has five values: 0, 0, 2, 1, and 0. This means that, during Month 5, there were warranty claims on 2 of the 1000 products sold in Month 3, and a warranty claim on 1 of the 1000 products sold in Month 4.

The warranty period for each product is 12 months, and the average cost for each warranty claim is $500. The reliability engineer wishes to predict warranty claims and costs for the next 5 months, with the assumption that the company will continue to sell 1000 products each month.

Recall from Chapters 3 and 4 that it is possible to determine the best fit of statistical distribution for a given set of failure time data. We will assume here that the failure times of the products follow Weibull distribution.

Section 6.1 illustrates the approach to convert the data from Table 6.1 into a format that Minitab will be able to process for warranty analysis. Section 6.2 gives the approach to perform warranty analysis. Section 6.3 gives an additional example with a different warranty period. Finally, Section 6.4 gives a concluding remark.

6.1 Preprocessing of Raw Data

Perform the following steps to convert the data from Table 6.1 into a format that Minitab will be able to process:

Step 1: Open a new Minitab project file and copy the data from Table 6.1, as shown in Figure 6.1.

TABLE 6.1

Sales and Warranty Claims for Sections 6.1 and 6.2

Sales	Month 1	Month 2	Month 3	Month 4	Month 5	Month 6	Month 7	Month 8	Month 9	Month 10	Month 11	Month 12
1000	0	0	0	0	0	0	0	0	0	0	0	0
1000		0	0	0	0	0	0	0	0	0	0	0
1000			0	0	2	4	3	1	1	2	4	5
1000				1	1	0	1	0	0	1	2	1
1000					0	0	0	1	3	1	4	2
1000						2	1	2	1	3	1	2
1000							1	0	1	1	2	1
1000								0	2	1	2	1
1000									0	0	1	0
1000										0	1	1
1000											1	0
1000												2

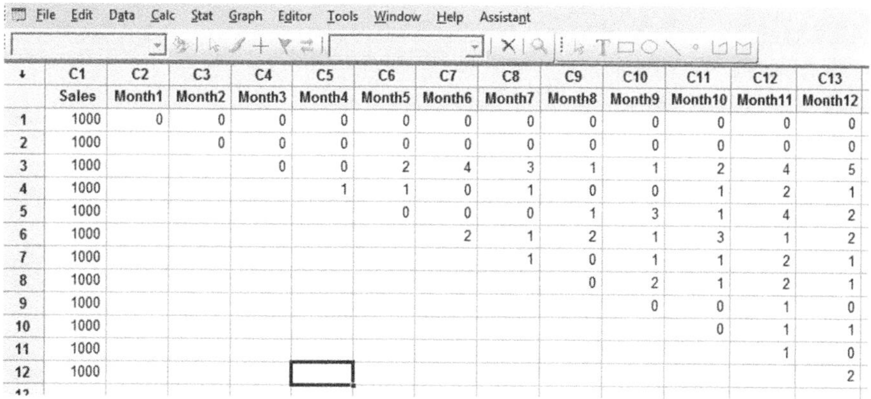

File Edit Data Calc Stat Graph Editor Tools Window Help Assistant

	C1	C2	C3	C4	C5	C6	C7	C8	C9	C10	C11	C12	C13
	Sales	Month1	Month2	Month3	Month4	Month5	Month6	Month7	Month8	Month9	Month10	Month11	Month12
1	1000	0	0	0	0	0	0	0	0	0	0	0	0
2	1000		0	0	0	0	0	0	0	0	0	0	0
3	1000			0	0	2	4	3	1	1	2	4	5
4	1000				1	1	0	1	0	0	1	2	1
5	1000					0	0	0	1	3	1	4	2
6	1000						2	1	2	1	3	1	2
7	1000							1	0	1	1	2	1
8	1000								0	2	1	2	1
9	1000									0	0	1	0
10	1000										0	1	1
11	1000											1	0
12	1000												2

FIGURE 6.1
Step 1 in Section 6.1.

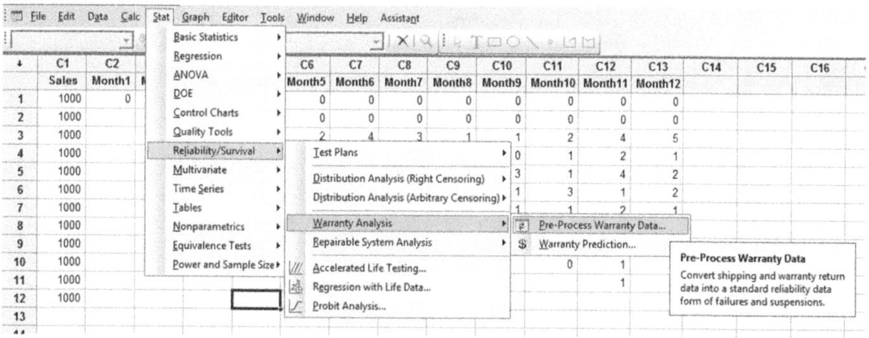

File Edit Data Calc Stat Graph Editor Tools Window Help Assistant

	C1	C2			C6	C7	C8	C9	C10	C11	C12	C13	C14	C15	C16
	Sales	Month1		Basic Statistics ▸	Month5	Month6	Month7	Month8	Month9	Month10	Month11	Month12			
1	1000	0		Regression ▸	0	0	0	0	0	0	0	0			
2	1000			ANOVA ▸	0	0	0	0	0	0	0	0			
3	1000			DOE ▸	2	4	3	1	1	2	4	5			
4	1000			Control Charts ▸	0	1	0	0	1	2	1				
5	1000			Quality Tools ▸	0	0	1	3	1	4	2				
6	1000			Reliability/Survival ▸	Test Plans ▸	2	1	3	1	2					
7	1000			Multivariate ▸		1	1	2	1						
8	1000			Time Series ▸	Distribution Analysis (Right Censoring) ▸	1	2	1							
9	1000			Tables ▸	Distribution Analysis (Arbitrary Censoring) ▸	1	2	1							
10	1000			Nonparametrics ▸	Warranty Analysis ▸	Pre-Process Warranty Data...									
11	1000			Equivalence Tests ▸	Repairable System Analysis ▸	Warranty Prediction...	0	1							
12	1000			Power and Sample Size ▸	Accelerated Life Testing...		1								
13					Regression with Life Data...										
					Probit Analysis...										

Pre-Process Warranty Data
Convert shipping and warranty return data into a standard reliability data form of failures and suspensions.

FIGURE 6.2
Step 2 commands in Section 6.1.

Step 2: As shown in Figure 6.2, go to *Stat → Reliability/Survival → Warranty Analysis → Pre-Process Warranty Data*. The dialog box shown in Figure 6.3 opens.

Step 3: As shown in Figure 6.4, select "Shipment values in a column" for *Data format*, select the "Sales" column for *Shipment (sale) column*, and select "Month 1" through to "Month 12" columns for *Return (failure) columns*.

Step 4: Click on *OK*, and you will see three new columns ("Start time", "End time", "Frequencies") in the worksheet, as shown in Figure 6.5.

It is important to note that, in Figure 6.5, a "Start time" of 0 and an "End time" of 1 means "during the first month after sale." Since there are 7 failures during the first month after sale, the frequency is 7. Similarly, a "Start time" of 1 and an "End time" of 2 means "during the second month after sale." Since there are 5 failures during the second month after sale, the frequency is 5, and so on.

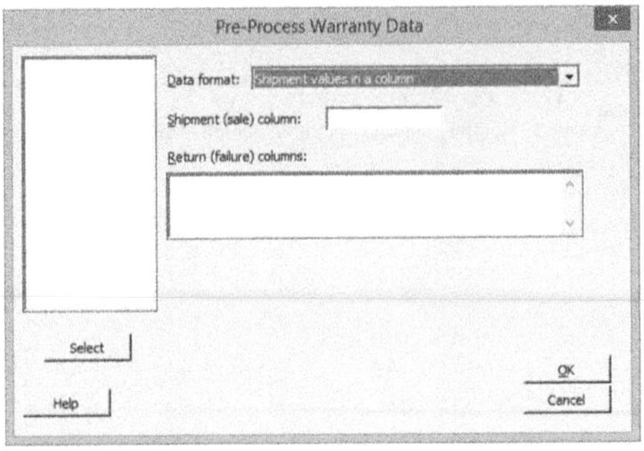

FIGURE 6.3
Step 2 dialog box in Section 6.1.

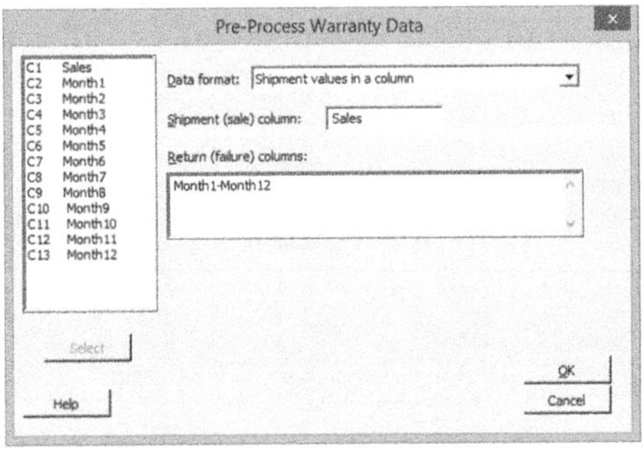

FIGURE 6.4
Step 3 in Section 6.1.

C2	C3	C4	C5	C6	C7	C8	C9	C10	C11	C12	C13	C14	C15	C16
Month1	Month2	Month3	Month4	Month5	Month6	Month7	Month8	Month9	Month10	Month11	Month12	Start time	End time	Frequencies
0	0	0	0	0	0	0	0	0	0	0	0	0	1	7
	0	0	0	0	0	0	0	0	0	0	0	1	2	5
		0	0	2	4	3	1	1	2	4	5	2	3	8
			1	1	0	1	0	0	1	2	1	3	4	10
				0	0	0	1	3	1	4	2	4	5	12
					2	1	2	1	3	1	2	5	6	4
						1	0	1	1	2	1	6	7	8
							0	2	1	2	1	7	8	6
								0	0	1	0	8	9	5
									0	1	1	9	10	5
										1	0	10	11	0
											2	11	12	0
												1	*	998
												2	*	999
												3	*	998
												4	*	999
												5	*	994
												6	*	994
												7	*	988
												8	*	989
												9	*	993
												10	*	978
												11	*	1000
												12	*	1000

FIGURE 6.5
Step 4 in Section 6.1.

Recall from Chapter 1 that a right-censored failure time means that the product that is being tested has not failed by that time. For example, if a product has not failed in, say, 100 hours, and if the test duration has ended, the failure time is recorded as 100 hours (right-censored). This product has not failed in 100 hours, but is expected to fail anytime later. The following examples explain how Minitab calculates the numbers in the "Frequencies" column for the right-censored failure times (Row 13 through to Row 24 in the worksheet):

Example 1: Of the 1000 products sold in the first month, none failed in any month up to the end of the 12th month after sale. Hence, at the end of 12 months ("Start time" of 12 and "End time" of "*"), 1000 units are still surviving. Hence, the respective frequency is 1000.

Example 2: Of the 1000 units sold in the second month, none failed in any month up to the end of the 11th month after sale. Hence, at the end of 11 months ("Start time" of 11 and "End time" of "*"), 1000 units are still surviving. Hence, the respective frequency is 1000.

Example 3: Of the 1000 units sold in the third month, 2 failed in Month 5, 4 failed in Month 6, 3 failed in Month 7, 1 failed in Month 8, 1 failed in Month 9, 2 failed in Month 10, 4 failed in Month 11, and 5 failed in Month 12. Hence, at the end of 10 months ("Start time" of 10 and "End time" of "*"), 978 units are still surviving. Hence, the respective frequency is 978.

6.2 Warranty Analysis

Perform the following steps for warranty analysis:

Step 1: As shown in Figure 6.6, go to *Stat* → *Reliability/Survival* → *Warranty Analysis* → *Warranty Prediction*. The dialog box shown in Figure 6.7 opens.

Step 2: As shown in Figure 6.8

- Select the "Start time" column for *Start time*.
- Select the "End time" column for *End time*.
- Select the "Frequencies" column for *Frequency (optional)*.
- Select "Weibull" for *Assumed distribution*.
- Enter "12" for *Length of warranty (optional)*.
- Enter "500" for *Average cost per failure (optional)*.

Step 3: Click on *Prediction*, and the dialog box shown in Figure 6.9 opens.

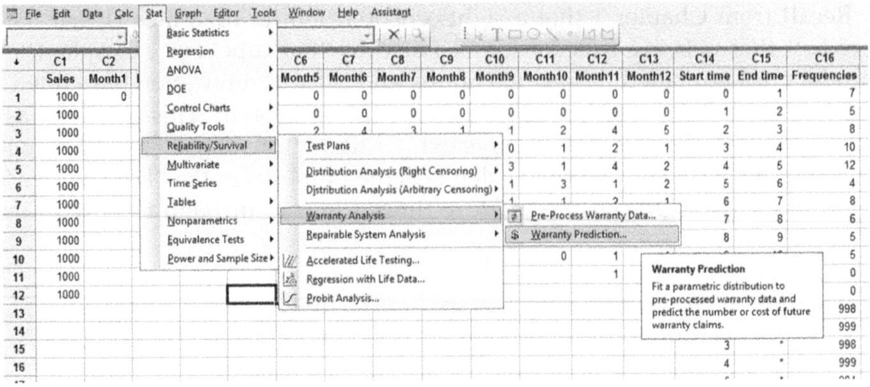

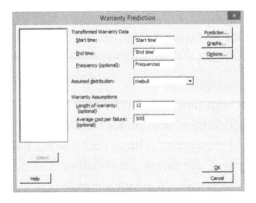

FIGURE 6.6
Step 1 commands in Section 6.2.

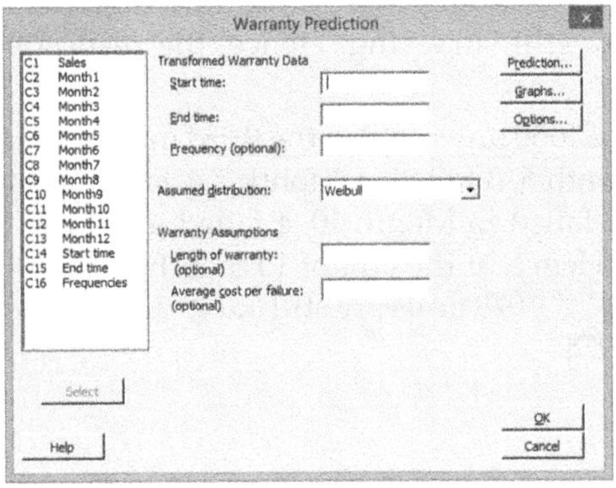

FIGURE 6.7
Step 1 dialog box in Section 6.2.

FIGURE 6.8
Step 2 in Section 6.2.

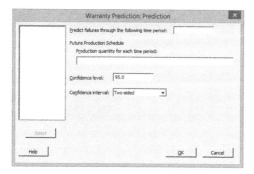

FIGURE 6.9
Step 3 in Section 6.2.

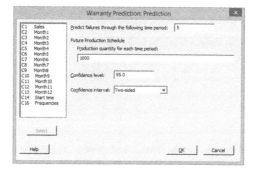

FIGURE 6.10
Step 4 in Section 6.2.

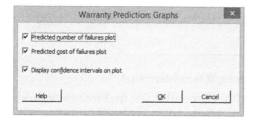

FIGURE 6.11
Step 6 in Section 6.2.

Step 4: As shown in Figure 6.10, enter "5" for *Predict failures through the following time period*, and enter "1000" for *Production quantity for each time period*.

Step 5: Click on *OK*, and it will take you back to the dialog box shown in Figure 6.8.

Step 6: Click on *Graphs*, and the dialog box shown in Figure 6.11 opens.

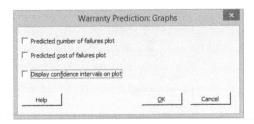

FIGURE 6.12
Step 7 in Section 6.2.

Step 7: As shown in Figure 6.12, uncheck all the boxes. (The respective plots are not useful for analysis in this case study.)

Step 8: Click on *OK*, and you will see the following output in the session window:

```
Number of units at risk for future time periods  10930
Warranty limit                                        12


Production Schedule

Future time period      1     2     3     4     5
Production quantity   1000  1000  1000  1000  1000


Table of Predicted Number of Failures and Cost

Future   Potential  Predicted                    Predicted
Time     Number of  Number of  95% Poisson CI    Cost of   95% Poisson CI
Period   Failures   Failures   Lower    Upper    Failures  Lower     Upper
  1        11930     11.9752   6.1828   20.9299    5987.6   3091.4   10465.0
  2        12930     23.9361  15.3262   35.6336   11968.0   7663.1   17816.8
  3        13930     35.9122  25.1405   49.7372   17956.1  12570.3   24868.6
  4        14930     47.8862  35.2938   63.5113   23943.1  17646.9   31755.7
  5        15930     59.8647  45.6682   77.0796   29932.4  22834.1   38539.8

Average cost per failure = 500
```

From this output, it is evident that

- At the end of 12 months, the first month of production goes out of warranty. Hence, the number of units at risk is 11,000 units (remaining) *minus* 70 units (units failed in the first month after sales, in Month 2 through to Month 12) = 10,930.
- In the next 1 month, 11,930 units (i.e., 10,930 + 1,000) are at risk. The number of failures predicted in the next 1 month is 11.9752. Hence, the expected cost of failure in the next 1 month is 11.9752 units * $500 per failure = $5,987.60.
- In the next 5 months, 15,930 units (i.e., 10,930 + 5,000) are at risk. The number of failures predicted in the next 5 months is 59.8647. Hence, the expected cost of failure in the next 5 months is 59.8647 units * $500 per failure = $29,932.40.

6.3 Additional Example

In this additional example, data regarding sales and warranty claims for the past 6 months have been collected, as shown in Table 6.2. The "Sales" column gives sales volume of 500 products each month. The columns labeled as "Month 1" through to "Month 6" show the warranty claims processed in each month. For example, the "Month 5" column has five values: 0, 0, 2, 1, and 0. This means that during Month 5, there were warranty claims on 2 of the 500 products sold in Month 3, and a warranty claim on 1 of the 500 products sold in Month 4.

The warranty period for each product is 6 months, and the average cost for each warranty claim is $1,000. The reliability engineer wishes to predict warranty claims and costs for the next 6 months, with the assumption that the company will continue to sell 500 products each month.

Recall from Chapters 3 and 4 that it is possible to determine the best fit of statistical distribution for a given set of failure time data. We will assume here that the failure times of the products follow the Weibull distribution.

Perform the following steps to convert the data from Table 6.2 into a format that Minitab will be able to process:

Step 1: Open a new Minitab project file and copy the data from Table 6.2, as shown in Figure 6.13.

TABLE 6.2

Sales and Warranty Claims for Section 6.3

Sales	Month 1	Month 2	Month 3	Month 4	Month 5	Month 6
500	0	0	0	0	0	0
500		0	0	0	0	0
500			0	0	2	4
500				1	1	0
500					0	0
500						2

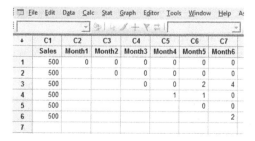

FIGURE 6.13
Step 1 for preprocessing data in Section 6.3.

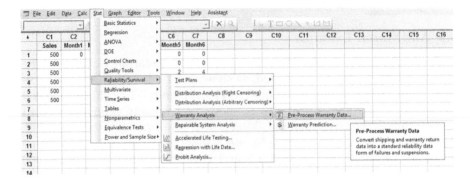

FIGURE 6.14
Step 2 commands for preprocessing data in Section 6.3.

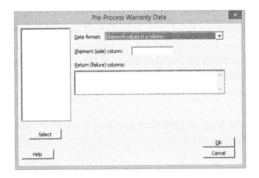

FIGURE 6.15
Step 2 dialog box for preprocessing data in Section 6.3.

Step 2: As shown in Figure 6.14, go to *Stat → Reliability/Survival → Warranty Analysis → Pre-Process Warranty Data*. The dialog box shown in Figure 6.15 opens.

Step 3: As shown in Figure 6.16, select "Shipment values in a column" for *Data format*, select the "Sales" column for *Shipment (sale) column*, and select "Month 1" through to "Month 6" columns for *Return (failure) columns*.

Step 4: Click on *OK*, and you will see three new columns ("Start time", "End time", "Frequencies") in the worksheet, as shown in Figure 6.17.

It is important to note that, in Figure 6.17, a "Start time" of 0 and an "End time" of 1 means "during the first month after sale." Since there are 3 failures during the first month after sale, the frequency is 3. Similarly, a "Start time" of 1 and an "End time" of 2 means "during the second month after sale." Since there is 1 failure during the second month after sale, the frequency is 1, and so on.

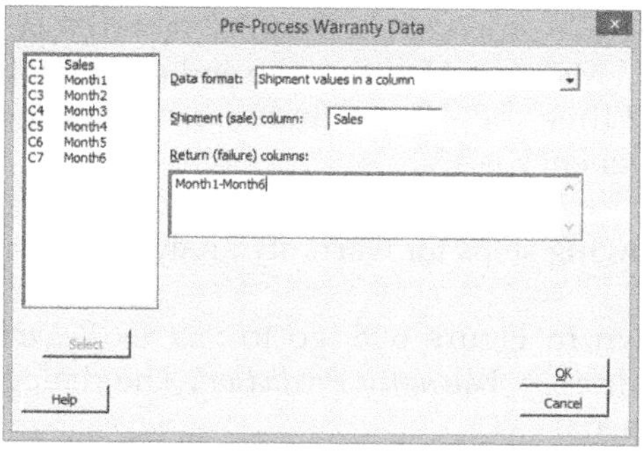

FIGURE 6.16
Step 3 for preprocessing data in Section 6.3.

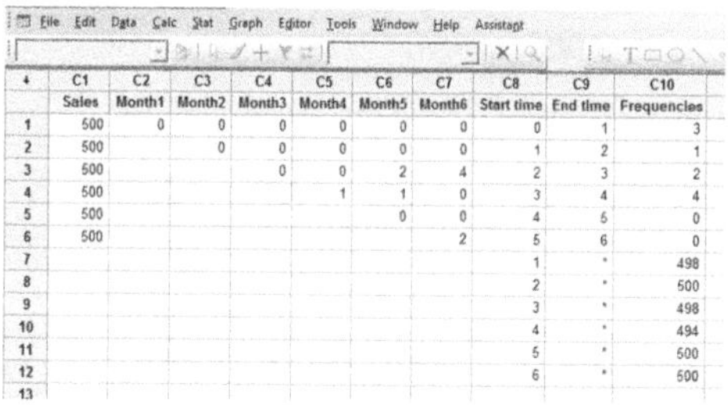

↓	C1	C2	C3	C4	C5	C6	C7	C8	C9	C10
	Sales	Month1	Month2	Month3	Month4	Month5	Month6	Start time	End time	Frequencies
1	500	0	0	0	0	0	0	0	1	3
2	500		0	0	0	0	0	1	2	1
3	500			0	0	2	4	2	3	2
4	500				1	1	0	3	4	4
5	500					0	0	4	5	0
6	500						2	5	6	0
7								1	*	498
8								2	*	500
9								3	*	498
10								4	*	494
11								5	*	500
12								6	*	500
13										

FIGURE 6.17
Step 4 for preprocessing data in Section 6.3.

Recall from Chapter 1 that a right-censored failure time means that the product that is being tested has not failed by that time. For example, if a product has not failed in, say, 100 hours, and if the test duration has ended, the failure time is recorded as 100 hours (right-censored). This product has not failed in 100 hours, but is expected to fail anytime later. The following three examples explain how Minitab calculates the numbers in the "Frequencies" column for the right-censored failure times (Row 7 through to Row 12 in the worksheet):

Example 1: Of the 500 products sold in the first month, none failed in any month up to the end of the sixth month after sale. Hence, at the end of 6 months ("Start time" of 6 and "End time" of "*"), 500 units are still surviving. Hence, the respective frequency is 500.

Example 2: Of the 500 units sold in the second month, none failed in any month up to the end of the fifth month after sale. Hence, at the end of 5 months ("Start time" of 5 and "End time" of "*"), 500 units are still surviving. Hence, the respective frequency is 500.

Example 3: Of the 500 units sold in the third month, 2 failed in Month 5, and 4 failed in Month 6. Hence, at the end of 4 months ("Start time" of 4 and "End time" of "*"), 494 units are still surviving. Hence, the respective frequency is 494.

Perform the following steps for warranty analysis:

Step 1: As shown in Figure 6.18, go to *Stat → Reliability/Survival → Warranty Analysis → Warranty Prediction*. The dialog box shown in Figure 6.19 opens.

Step 2: As shown in Figure 6.20

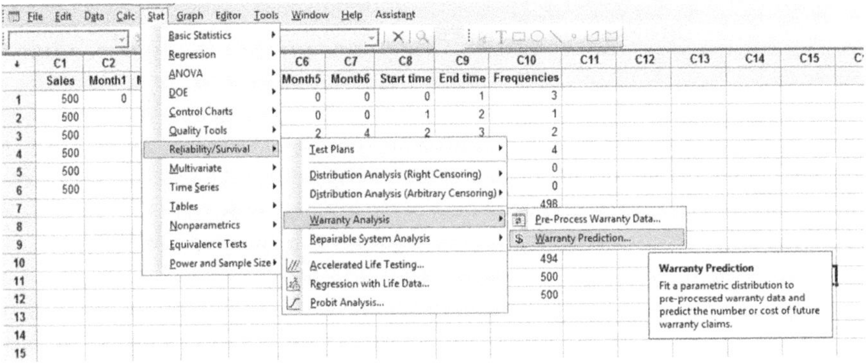

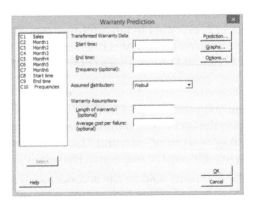

FIGURE 6.18
Step 1 commands for warranty analysis in Section 6.3.

FIGURE 6.19
Step 1 dialog box for warranty analysis in Section 6.3.

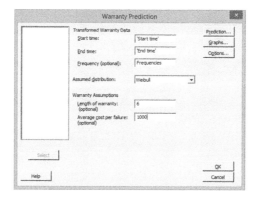

FIGURE 6.20
Step 2 for warranty analysis in Section 6.3.

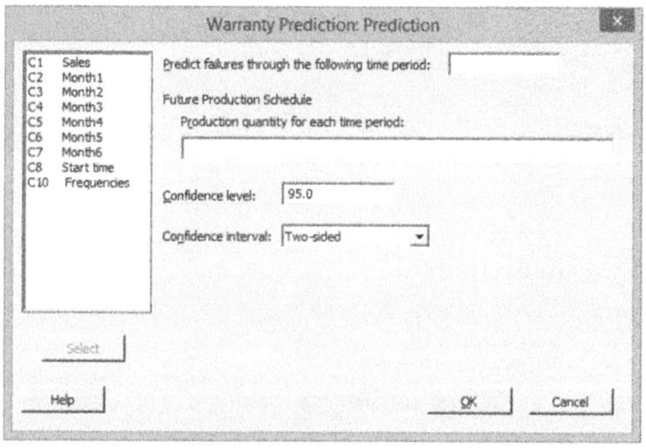

FIGURE 6.21
Step 3 for warranty analysis in Section 6.3.

- Select the "Start time" column for *Start time.*
- Select the "End time" column for *End time.*
- Select the "Frequencies" column for *Frequency (optional).*
- Select "Weibull" for *Assumed distribution.*
- Enter "6" for *Length of warranty (optional).*
- Enter "1000" for *Average cost per failure (optional).*

Step 3: Click on *Prediction,* and the dialog box shown in Figure 6.21 opens.

Step 4: As shown in Figure 6.22, enter "6" for *Predict failures through the following time period* and enter "500" or *Production quantity for each time period.*

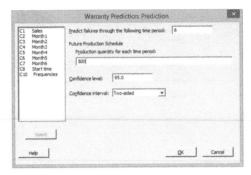

FIGURE 6.22
Step 4 for warranty analysis in Section 6.3.

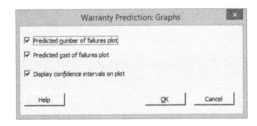

FIGURE 6.23
Step 6 for warranty analysis in Section 6.3.

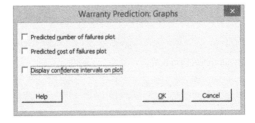

FIGURE 6.24
Step 7 for warranty analysis in Section 6.3.

Step 5: Click on *OK*, and it will take you back to the dialog box shown in Figure 6.20.

Step 6: Click on *Graphs*, and the dialog box shown in Figure 6.23 opens.

Step 7: As shown in Figure 6.24, uncheck all the boxes. (The respective plots are not useful for analysis in this case study.)

Step 8: Click on *OK*, and you will see the following output in the session window:

```
Number of units at risk for future time periods  2490
Warranty limit                                       6


Production Schedule

Future time period     1    2    3    4    5    6
Production quantity   500  500  500  500  500  500


Table of Predicted Number of Failures and Cost

Future  Potential  Predicted                    Predicted
 Time   Number of  Number of   95% Poisson CI    Cost of   95% Poisson CI
Period   Failures   Failures  Lower    Upper    Failures  Lower     Upper
  1        2990      2.9004   0.5762   8.6172     2900.4    576.2    8617.2
  2        3490      5.7984   2.0823  12.7815     5798.4   2082.3   12781.5
  3        3990      8.7005   3.9151  16.6908     8700.5   3915.1   16690.8
  4        4490     11.6031   5.9175  20.4551    11603.1   5917.5   20455.1
  5        4990     14.5049   8.0271  24.1220    14504.9   8027.1   24122.0
  6        5490     17.4081  10.2145  27.7211    17408.1  10214.5   27721.1

Average cost per failure = 1000
```

From this output, it is evident that

- At the end of 6 months, the first month of production goes out of warranty. Hence, the number of units at risk is 2,500 units (remaining) *minus* 10 units (units failed in the first month after sales, in Month 2 through to Month 6) = 2,490.
- In the next 1 month, 2,990 units (i.e., 2,490 + 500) are at risk. The number of failures predicted in the next 1 month is 2.9004. Hence, the expected cost of failure in the next 1 month is 2.9004 units * $1,000 per failure = $2,900.40.
- In the next 6 months, 5,490 units (i.e., 2,490 + 3,000) are at risk. The number of failures predicted in the next 5 months is 17.4081. Hence, the expected cost of failure in the next 6 months is 17.4081 units * $1,000 per failure = $17,408.10.

6.4 Conclusion

These case studies illustrated how to use the *Warranty Analysis* option in Minitab to predict the amount of money that is needed to cover the warranty costs for units in a specific period of time in the future.

7

Accelerated Life Testing

In situations with long product lead times, manufacturers prefer reducing the test times of products by testing them under, say, abnormally high temperatures, which may lead to quicker product failures and, in turn, shorter product lead times. This is called *accelerated life testing*, and is especially useful when the relationship between the test temperature and product failure time is known. A relationship that is commonly used in the plastic and semiconductor industries is the Arrhenius relationship. The concepts of physics and chemistry, which are needed to understand the Arrhenius relationship, are beyond the scope of this book. Readers interested in those concepts and the respective mathematical formulas are encouraged to review the book listed in the Bibliography.

In this case study, a reliability engineer wishes to perform accelerated life testing on a plastic product. Section 7.1 uses Minitab® to illustrate the analysis of test results. Section 7.2 provides an additional example. Finally, Section 7.3 gives a concluding remark.

7.1 Accelerated Life Testing and Analysis

The normal operating temperatures of the product range from 80°C to 100°C. However, to save time, the reliability engineer performs accelerated life testing on a random sample of 80 products, at the following abnormally high temperatures: 110°C, 130°C, 150°C, and 170°C. The engineer is interested in knowing the times by which 50%, 90%, and 99% of the products are expected to fail, at 80°C and 100°C.

The failure times (in hours) are recorded in Table 7.1. In the "Censor" column of the table, "C" refers to right-censored failure time and "F" refers to exact failure time. Recall from Chapter 1 that a right-censored failure time means that the product that is being tested has not failed by that time. For example, if a product has not failed in, say, 100 hours, and if the test duration has ended, the failure time is recorded as 100 hours (right-censored). This product has not failed in 100 hours, but is expected to fail anytime later. Recall also that an exact failure time means that the product has failed exactly at that time. For example, if a product has failed exactly at the end of, say, 50 hours, the failure time is recorded as 50 hours (exact).

TABLE 7.1

Failure Time Data for Section 7.1

No.	Temp	Failure Time	Censor
1	170	345	F
2	170	870	F
3	170	244	C
4	170	716	F
5	170	531	F
6	170	738	F
7	170	461	F
8	170	221	F
9	170	665	F
10	170	384	C
11	170	394	C
12	170	369	F
13	170	366	F
14	170	507	F
15	170	461	F
16	170	431	F
17	170	479	F
18	170	106	F
19	170	545	F
20	170	536	F
21	150	2,134	C
22	150	2,746	F
23	150	2,859	F
24	150	1,826	C
25	150	996	F
26	150	2,733	F
27	150	3,651	F
28	150	2,073	F
29	150	2,291	F
30	150	1,689	F
31	150	1,533	F
32	150	1,752	F
33	150	1,764	F
34	150	2,042	F
35	150	1,043	C
36	150	1,214	F
37	150	3,154	F
38	150	2,386	F
39	150	2,190	F
40	150	1,642	F
41	130	8,290	F

TABLE 7.1 (CONTINUED)

Failure Time Data for Section 7.1

No.	Temp	Failure Time	Censor
42	130	10,183	F
43	130	3,987	F
44	130	3,545	F
45	130	4,735	C
46	130	7,919	F
47	130	4,925	F
48	130	2,214	F
49	130	5,351	F
50	130	3,147	F
51	130	7,304	F
52	130	6,947	F
53	130	5,355	F
54	130	3,308	F
55	130	4,373	F
56	130	6,226	F
57	130	5,117	F
58	130	3,620	C
59	130	3,128	F
60	130	4,348	F
61	110	21,900	C
62	110	13,218	F
63	110	17,610	F
64	110	7,336	C
65	110	18,397	F
66	110	13,673	F
67	110	8,702	F
68	110	21,900	C
69	110	13,513	F
70	110	14,482	F
71	110	20,975	F
72	110	12,090	F
73	110	17,822	F
74	110	11,769	F
75	110	21,900	C
76	110	16,289	F
77	110	21,900	C
78	110	18,806	F
79	110	11,142	C
80	110	17,782	F

It is known that the Arrhenius relationship exists between the temperature and failure time of the product. Also, it has been established that failure times follow Weibull distribution. Recall from Chapter 3 how to determine which statistical distribution is the best fit for failure time data.

Perform the following steps for accelerated life testing of the product:

Step 1: Open a new Minitab project file and copy the data from Table 7.1, as shown in Figure 7.1. Also, create a Design column and enter the normal operating temperature limits of 80°C and 100°C.

Step 2: As shown in Figure 7.2, go to *Stat → Reliability/Survival → Accelerated Life Testing*. The dialog box shown in Figure 7.3 opens.

	C1	C2	C3-T	C4	
	Temp	Failure Time	Censor	Design	
1	170	345	F	80	
2	170	870	F	100	
3	170	244	C		
4	170	716	F		
5	170	531	F		
6	170	738	F		
7	170	461	F		
8	170	221	F		
9	170	665	F		
10	170	384	C		
11	170	394	C		
12	170	369	F		
13	170	366	F		
14	170	507	F		
15	170	461	F		
16	170	431	F		
17	170	479	F		

FIGURE 7.1
Step 1 in Section 7.1.

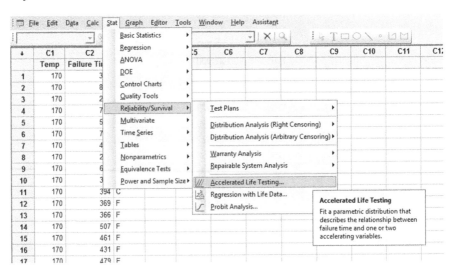

FIGURE 7.2
Step 2 commands in Section 7.1.

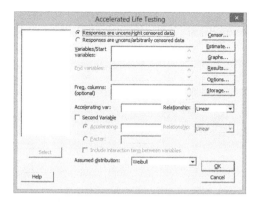

FIGURE 7.3
Step 2 dialog box in Section 7.1.

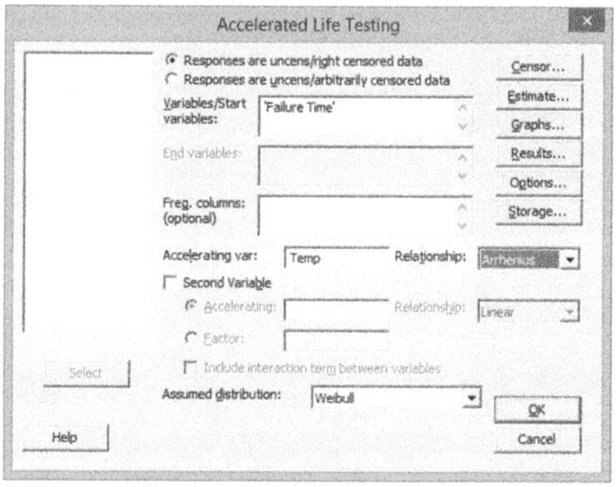

FIGURE 7.4
Step 3 in Section 7.1.

Step 3: As shown in Figure 7.4, select *Responses are uncens/right censored data*, select the "Failure Time" column for *Variables/Start variables*, select the "Temp" column for *Accelerating var*, select "Arrhenius" for *Relationship*, and select "Weibull" for *Assumed distribution*.

Step 4: Click on *Censor*, and the dialog box shown in Figure 7.5 opens.

Step 5: As shown in Figure 7.6, select the "Censor" column for *Use censoring columns* and enter "C" in *Censoring value*.

Step 6: Click on *OK*, and it will take you back to the dialog box shown in Figure 7.4.

Step 7: Click on *Estimate*, and the dialog box shown in Figure 7.7 opens.

Step 8: As shown in Figure 7.8, select the "Design" column for *Enter new predictor values* and enter "50 90 99" in *Estimate percentiles for percents*.

Step 9: Click on *OK*, and it will take you back to the dialog box shown in Figure 7.4.

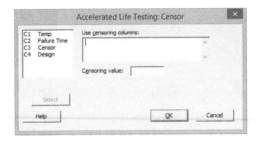

FIGURE 7.5
Step 4 in Section 7.1.

FIGURE 7.6
Step 5 in Section 7.1.

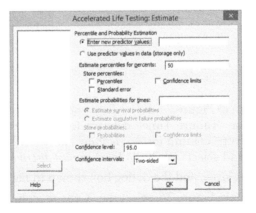

FIGURE 7.7
Step 7 in Section 7.1.

Step 10: Click on *Graphs*, and the dialog box shown in Figure 7.9 opens.

Step 11: As shown in Figure 7.10, uncheck the boxes for *Relation plot* and *Probability plot for each accelerating level based on fitted model.* (These plots are not useful in this case study.)

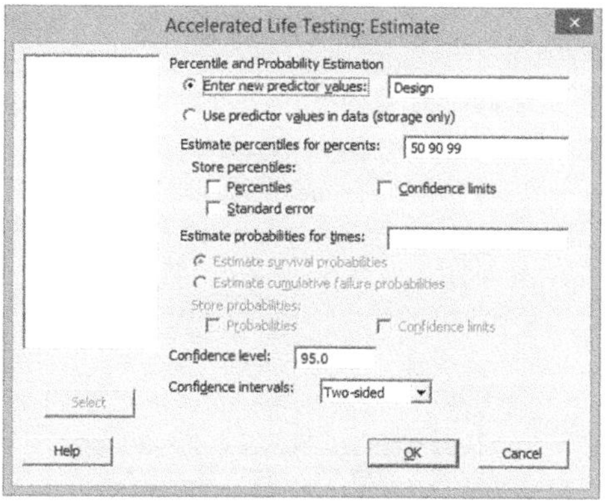

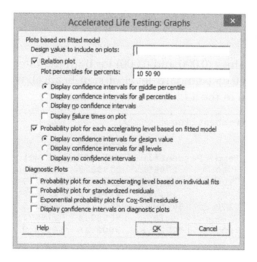

FIGURE 7.8
Step 8 in Section 7.1.

FIGURE 7.9
Step 10 in Section 7.1.

Step 12: Click on *OK*, and it will take you back to the dialog box shown in Figure 7.4.

Step 13: Click on *OK*, and you will see the following output in the session window:

Regression Table						
		Standard			95.0% Normal CI	
Predictor	Coef	Error	Z	P	Lower	Upper
Intercept	−15.1857	0.985965	−15.40	0.000	−17.1182	−13.2533
Temp	0.830664	0.0350341	23.71	0.000	0.761998	0.899329
Shape	2.82501	0.256998			2.36365	3.37641

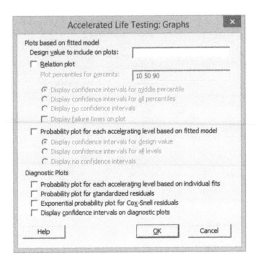

FIGURE 7.10
Step 11 in Section 7.1.

Since the *p*-values (0.000 and 0.000) for both the Intercept and Temp are less than 0.05 (standard level of significance), we can trust the output shown in the additional session window:

```
Table of Percentiles

                                 Standard     95.0% Normal CI
Percent   Temp   Percentile        Error     Lower      Upper
     50     80       159552       27435.3    113903     223495
     50    100      36944.8       4215.21   29541.7    46203.1
     90     80       244041       42847.8    172987     344282
     90    100      56508.7       6626.56   44905.3    71110.3
     99     80       311905       56587.4    218571     445094
     99    100      72222.8       9002.15   56568.8    92208.6
```

Based on this output, it is evident that

- 50% of the products are expected to fail in 159,552 hours at 80°C.

- 50% of the products are expected to fail in 36,944.8 hours at 100°C.

- 90% of the products are expected to fail in 244,041 hours at 80°C.

- 90% of the products are expected to fail in 56,508.7 hours at 100°C.

- 99% of the products are expected to fail in 311,905 hours at 80°C.

- 99% of the products are expected to fail in 72,222.8 hours at 100°C.

7.2 Additional Example

In this example, the normal operating temperatures of the product range from 100°C to 130°C. However, to save time, the reliability engineer performs accelerated life testing on a random sample of 15 products, at the following abnormally high temperatures: 200°C, 220°C, and 230°C. The engineer is interested in knowing the times by which 50%, 90%, and 99% of the products are expected to fail, at 100°C and 130°C.

The failure times (in days) are recorded in Table 7.2. In the "Censor" column of the table, "C" refers to right-censored failure time and "F" refers to exact failure time.

It is known that the Arrhenius relationship exists between the temperature and failure time of the product. Also, it has been established that failure times follow Weibull distribution.

Perform the following steps for accelerated life testing of the product:

Step 1: Open a new Minitab project file and copy the data from Table 7.2, as shown in Figure 7.11. Also, create a Design column and enter the normal operating temperature limits of 100°C and 130°C.

Step 2: As shown in Figure 7.12, go to *Stat → Reliability/Survival → Accelerated Life Testing*. The dialog box shown in Figure 7.13 opens.

Step 3: As shown in Figure 7.14, select *Responses are uncens/right censored data*, select the "Failure Time (days)" column for *Variables/Start variables*,

TABLE 7.2

Failure Time Data for Section 7.2

No.	Temp	Failure Time	Censor
1	200	120	F
2	200	110	F
3	200	98	C
4	200	101	F
5	200	102	F
6	220	90	F
7	220	89	F
8	220	91	F
9	220	85	F
10	220	88	C
11	230	65	C
12	230	64	F
13	230	69	F
14	230	63	F
15	230	65	F

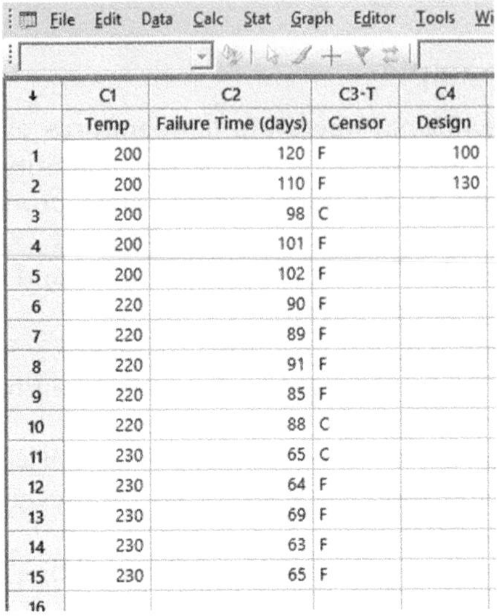

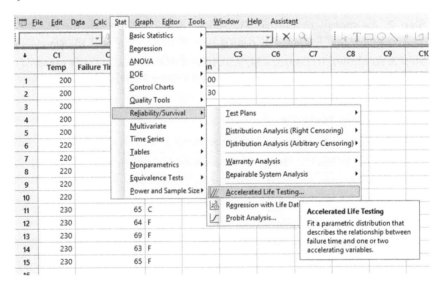

FIGURE 7.11
Step 1 in Section 7.2.

FIGURE 7.12
Step 2 commands in Section 7.2.

select the "Temp" column for *Accelerating var*, select "Arrhenius" for *Relationship*, and select "Weibull" for *Assumed distribution*.

Step 4: Click on *Censor*, and the dialog box shown in Figure 7.15 opens.

Step 5: As shown in Figure 7.16, select the "Censor" column for *Use censoring columns* and enter "C" in *Censoring value*.

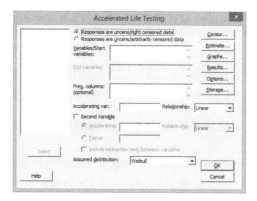

FIGURE 7.13
Step 2 dialog box in Section 7.2.

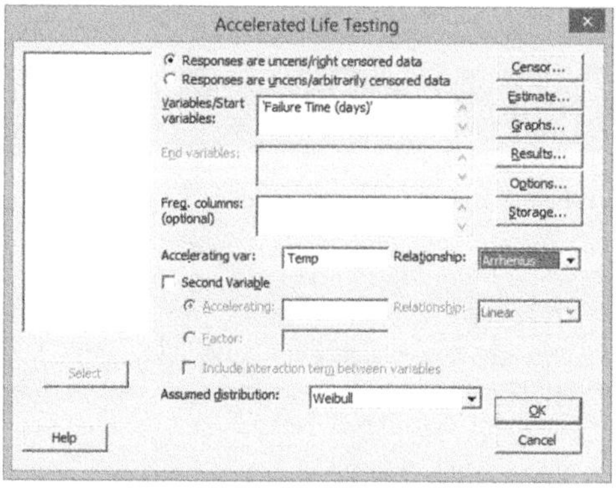

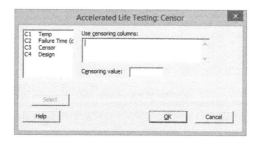

FIGURE 7.14
Step 3 in Section 7.2.

FIGURE 7.15
Step 4 in Section 7.2.

Step 6: Click on *OK*, and it will take you back to the dialog box shown in Figure 7.14.

Step 7: Click on *Estimate*, and the dialog box shown in Figure 7.17 opens.

Step 8: As shown in Figure 7.18, select the "Design" column for *Enter new predictor values* and enter "50 90 99" in *Estimate percentiles for percents*.

Step 9: Click on *OK*, and it will take you back to the dialog box shown in Figure 7.14.

Step 10: Click on *Graphs*, and the dialog box shown in Figure 7.19 opens.

Step 11: As shown in Figure 7.20, uncheck the boxes for *Relation plot* and *Probability plot for each accelerating level based on fitted model*. (These plots are not useful in this case study.)

Step 12: Click on *OK*, and it will take you back to the dialog box shown in Figure 7.14.

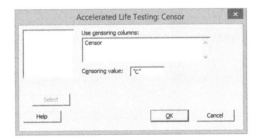

FIGURE 7.16
Step 5 in Section 7.2.

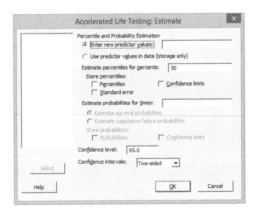

FIGURE 7.17
Step 7 in Section 7.2.

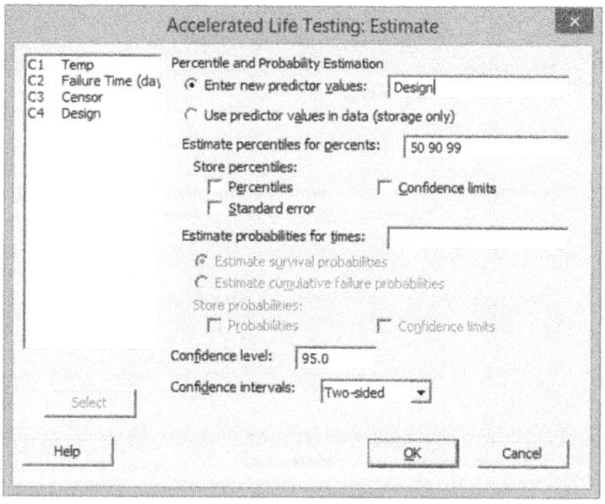

FIGURE 7.18
Step 8 in Section 7.2.

FIGURE 7.19
Step 10 in Section 7.2.

Step 13: Click on *OK*, and you will see the following output in the session window:

```
Regression Table

                     Standard                      95.0% Normal CI
Predictor     Coef     Error      Z        P      Lower      Upper
Intercept  -2.71940   1.11909   -2.43    0.015   -4.91278  -0.526027
Temp       0.304619  0.0471903   6.46    0.000   0.212128   0.397110
Shape      13.7945    3.12835                     8.84439   21.5151
```

Since the *p*-values (0.015 and 0.000) for both the Intercept and Temp are less than 0.05 (standard level of significance), we can trust the output shown in the additional session window:

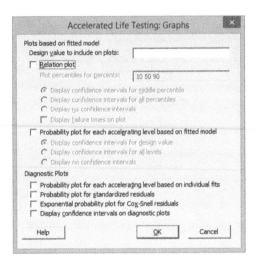

FIGURE 7.20
Step 11 in Section 7.2.

```
Table of Percentiles

                                  Standard    95.0% Normal CI
Percent   Temp   Percentile      Error     Lower      Upper
     50    100      834.896     291.595   421.059   1655.47
     50    130      412.559     99.1825   257.544   660.879
     90    100      910.813     318.667   458.795   1808.17
     90    130      450.074     108.472   280.631   721.823
     99    100      957.749     335.911   481.626   1904.56
     99    130      473.267     114.573   294.469   760.629
```

Based on this output, it is evident that

- 50% of the products are expected to fail in 834.896 days at 100°C.
- 50% of the products are expected to fail in 412.559 days at 130°C.
- 90% of the products are expected to fail in 910.813 days at 100°C.
- 90% of the products are expected to fail in 450.074 days at 130°C.
- 99% of the products are expected to fail in 957.749 days at 100°C.
- 99% of the products are expected to fail in 473.267 days at 130°C.

7.3 Conclusion

These case studies illustrated how to use Minitab to analyze the results from accelerated life testing on two different products.

8

Reliability Test Plan with Constrained Test Time and Number of Failures

Due to schedule and cost constraints, reliability tests are conducted on a sample of products instead of the entire population. Many a time, it is prohibitively expensive or time consuming or both to test the sample for a large period of time. Also, the company may not be willing to allow a large number of failures in the test, especially if the product is of high value. Amid such constraints, the reliability engineer is responsible for determining the sample size required for the test. This case study illustrates how to determine the test sample size using Minitab®, when the testing time and the number of failures are constrained.

A reliability engineer at an electrical motor manufacturing firm wishes to calculate the test sample size that is required to assess the reliability of a new motor, which is about to be launched in the market. The historical reliability standard for electrical motors produced by this firm has been "at least 90% survival after 1000 hours." The firm has conducted prior studies on similar motors, and believes that the "time to failure" distribution for the new motor is Weibull, with a shape parameter of 3.

Given the schedule and cost constraints, the reliability engineer can afford to run the demonstration for 800 hours, and wants the experiment to result in no more than two failures. The engineer wishes to know the number of motors he must test for 800 hours and also the results he must show, in order to claim that the new motor satisfies the historical standard of "at least 90% survival after 1000 hours."

Section 8.1 illustrates the approach to determine the test sample size using Minitab for the historical standard of "at least 90% survival after 1000 hours," with a test time of 800 hours and a maximum allowed number of failures of 2. Section 8.2 gives a few additional examples for different constraints and also compares the respective outputs. Finally, Section 8.3 gives a concluding remark.

8.1 Determination of Test Sample Size

Perform the following steps to determine the test sample size:

Step 1: Open a new Minitab project file and go to *Demonstration*, as shown in Figure 8.1.

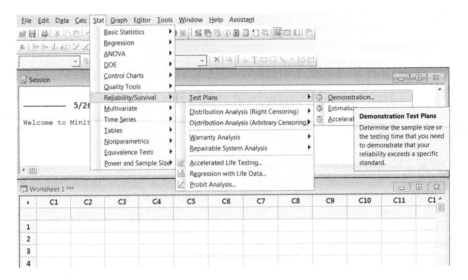

FIGURE 8.1
Step 1 in Section 8.1.

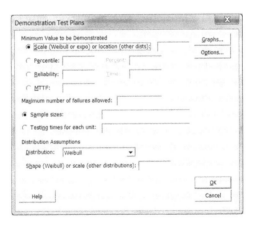

FIGURE 8.2
Step 2 in Section 8.1.

Step 2: The dialog box shown in Figure 8.2 opens.

Step 3: As shown in Figure 8.3, enter

- "0.90" for *Reliability*
- "1000" for *Time*
- "2" for *Maximum number of failures allowed*
- "800" for *Testing times for each unit*
- "Weibull" for *Distribution*
- "3" for *Shape (Weibull) or scale (other distributions)*

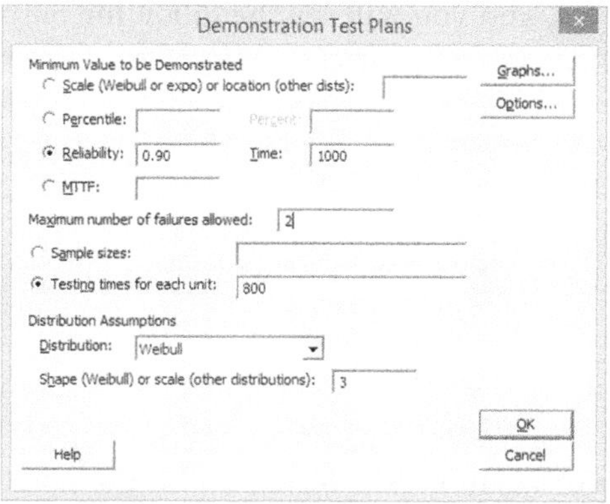

FIGURE 8.3
Step 3 in Section 8.1.

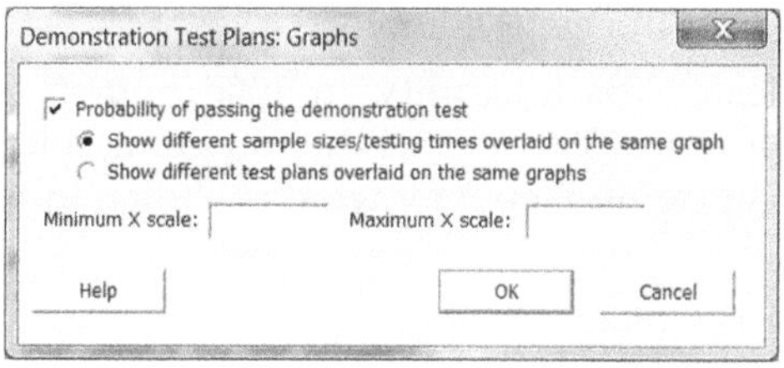

FIGURE 8.4
Step 4 in Section 8.1.

Step 4: Click on *Graphs,* and the dialog box shown in Figure 8.4 opens.

Step 5: As shown in Figure 8.5, uncheck the box for *Probability of passing the demonstration test.* (We are unchecking this box because it does not help to address the engineer's purpose in this case.)

Step 6: Click on *OK,* and it will take you back to the dialog box shown in Figure 8.3.

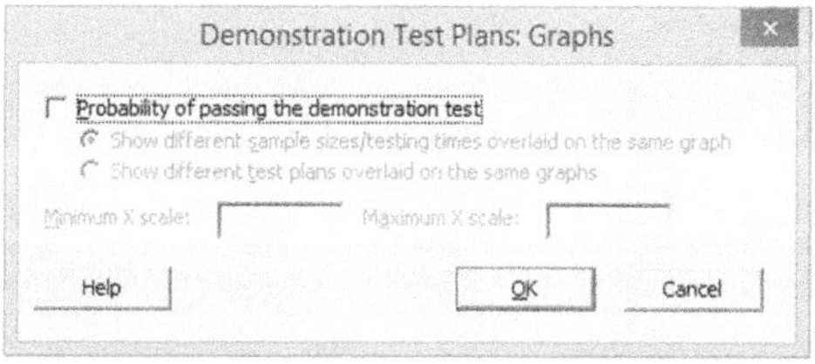

FIGURE 8.5
Step 5 in Section 8.1.

Step 7: Click on *OK*, and you will see the following output:

```
Reliability Test Plan
Distribution: Weibull, Shape = 3
Reliability Goal = 0.9, Target Confidence Level = 95%


                                  Actual
Failure   Testing   Sample    Confidence
   Test      Time     Size         Level
      2       800      118       95.0557
```

From this output, the reliability engineer can claim the following with 95.0557% confidence:

If a sample of 118 motors is tested and if no more than 2 of them fail in 800 hours, it means that the historical standard of "at least 90% reliability in 1000 hours" is achieved for all units.

Notice that Minitab had to compromise on the round figure of the confidence level (95.0557%) because it had to round the test sample size to the nearest integer (118 motors).

Since "at least 90% reliability in 1000 hours" also means "at most 10% failure in 1000 hours," the foregoing output can also be obtained by replacing Step 3 by what is shown in Figure 8.6. That is, enter

- "1000" for *Percentile*
- "10" for *Percent*
- "2" for *Maximum number of failures allowed*
- "800" for *Testing times for each unit*
- "Weibull" for *Distribution*
- "3" for *Shape (Weibull) or scale (other distributions)*

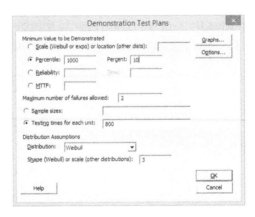

FIGURE 8.6
Alternative in Section 8.1.

8.2 Additional Scenarios

The reliability engineer now wishes to know what the sample size would be if he can afford to run the experiment for more than 800 hours, for the same number of maximum allowed failures, that is, 2. To this end, he replaces Step 3 in Section 8.1 by performing what are shown in Figures 8.7 through 8.9, one at a time.

The output for 1000 hours (Figure 8.7) of test time is as follows:

```
Reliability Test Plan
Distribution: Weibull, Shape = 3
Reliability Goal = 0.9, Target Confidence Level = 95%


                              Actual
Failure  Testing  Sample   Confidence
   Test     Time    Size       Level
      2     1000      61     95.0882
```

The output for 1500 hours (Figure 8.8) of test time is as follows:

```
Reliability Test Plan
Distribution: Weibull, Shape = 3
Reliability Goal = 0.9, Target Confidence Level = 95%


                              Actual
Failure  Testing  Sample   Confidence
   Test     Time    Size       Level
      2     1500      19     95.3113
```

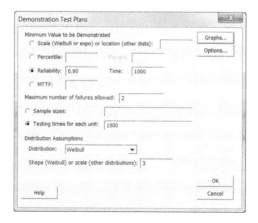

FIGURE 8.7
Test plan for 1000 hours in Section 8.2.

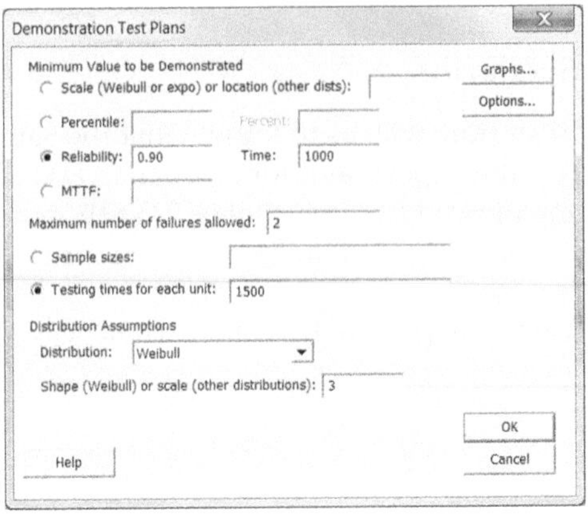

FIGURE 8.8
Test plan for 1500 hours in Section 8.2.

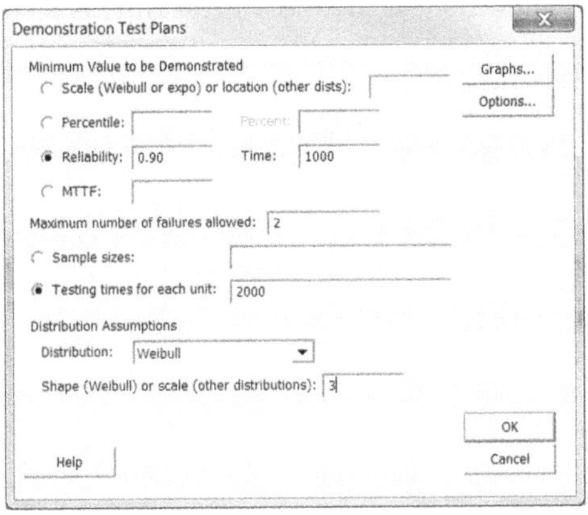

FIGURE 8.9
Test plan for 2000 hours in Section 8.2.

The output for 2000 hours (Figure 8.9) of test time is as follows:

```
Reliability Test Plan
Distribution: Weibull, Shape = 3
Reliability Goal = 0.9, Target Confidence Level = 95%


                                   Actual
Failure   Testing   Sample   Confidence
  Test      Time     Size       Level
     2      2000        9      96.1466
```

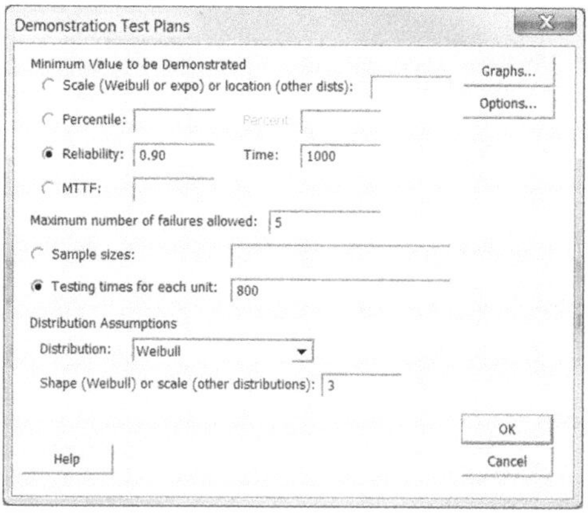

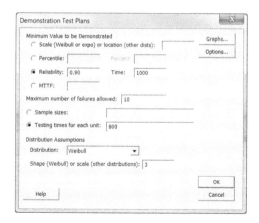

FIGURE 8.10
Test plan for five allowed failures in Section 8.2.

It is interesting to note from these outputs that for a fixed number of maximum allowed failures, as the testing time increases, the required test sample size reduces. In other words, the longer we test the sample, the smaller the required size of the sample is. For example, the test sample size reduces from 19 to 9, as the testing time increases from 1500 to 2000 hours.

The reliability engineer then wishes to know what the sample size would be if he can afford more than two failures, for the same testing time of 800 hours. To this end, he replaces Step 3 in Section 8.1 by performing what is shown in Figures 8.10 through 8.12, one at a time.

FIGURE 8.11
Test plan for 10 allowed failures in Section 8.2.

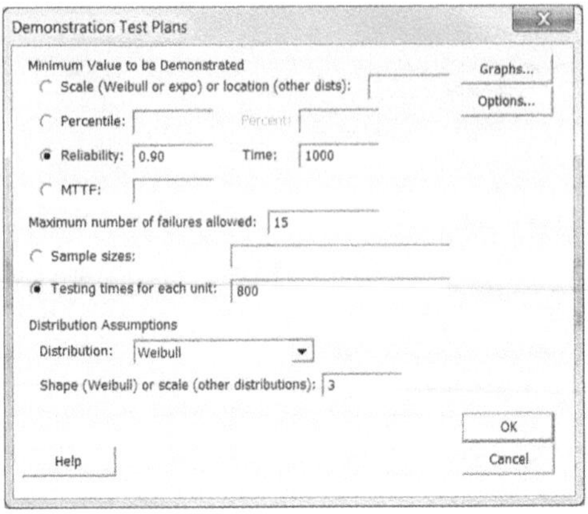

FIGURE 8.12
Test plan for 15 allowed failures in Section 8.2.

The output for five allowed failures (Figure 8.10) is as follows:

```
Reliability Test Plan
Distribution: Weibull, Shape = 3
Reliability Goal = 0.9, Target Confidence Level = 95%


                                Actual
Failure  Testing  Sample    Confidence
  Test     Time     Size        Level
     5      800      197      94.9362
```

The output for 10 allowed failures (Figure 8.11) is as follows:

```
Reliability Test Plan
Distribution: Weibull, Shape = 3
Reliability Goal = 0.9, Target Confidence Level = 95%


                                Actual
Failure  Testing  Sample    Confidence
  Test     Time     Size        Level
    10      800      319      94.9397
```

The output for 15 allowed failures (Figure 8.12) is as follows:

```
Reliability Test Plan
Distribution: Weibull, Shape = 3
Reliability Goal = 0.9, Target Confidence Level = 95%


                                Actual
Failure  Testing  Sample    Confidence
  Test     Time     Size        Level
    15      800      436      95.0302
```

It is interesting to note from these outputs that for a fixed testing time, as the maximum allowed number of failures increases, the required test sample size increases as well. In other words, to accommodate for more (potential) failures, the engineer is required to test more products. For example, the test sample size increases from 319 to 436, as the maximum allowed number of failures increases from 10 failures to 15 failures. An interesting question for the reader to think about is: What is the engineer's motivation for increasing the maximum allowed number of failures if it requires him to increase the sample size and hence increase the cost. The answer lies in the "power" of the demonstration test. The "power" gives us an indication of how reliable the test is. Suppose that the engineer wishes to compare the "power" of the tests for 2 maximum allowed failures and 25 maximum allowed failures. Perform the following steps:

Step 1: Open a new Minitab project file and go to *Demonstration*, as shown in Figure 8.1.

Step 2: The dialog box shown in Figure 8.2 opens.

Step 3: As shown in Figure 8.13, enter

- "0.90" for *Reliability*
- "1000" for *Time*
- "2, 25" for *Maximum number of failures allowed*
- "800" for *Testing times for each unit*
- "Weibull" for *Distribution*
- "3" for *Shape (Weibull) or scale (other distributions)*

Step 4: Click on *Graphs*, and the dialog box shown in Figure 8.4 opens.

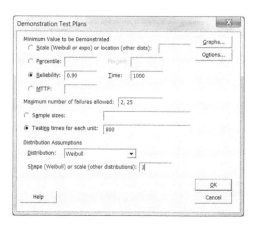

FIGURE 8.13
Step 3 in Section 8.2.

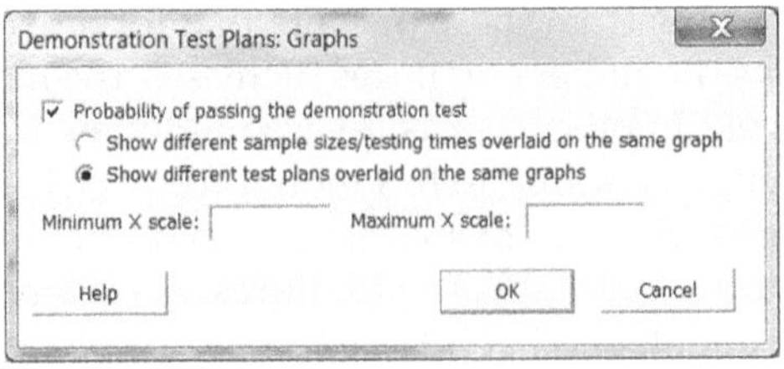

Demonstration Test Plans: Graphs   X

☑ Probability of passing the demonstration test
  ○ Show different sample sizes/testing times overlaid on the same graph
  ● Show different test plans overlaid on the same graphs

Minimum X scale: [   ]   Maximum X scale: [   ]

    Help   OK   Cancel

FIGURE 8.14
Step 5 in Section 8.2.

Step 5: As shown in Figure 8.14, select *Show different test plans overlaid on the same graphs.*

Step 6: Click on *OK,* and it will take you back to the dialog box shown in Figure 8.13.

Step 7: Click on *OK,* and you will see the graph shown in Figure 8.15.

Recall that the reliability standard in this case study is "at least 90% survival after 1000 hours." If the true (unknown) performance of the motors is, say, "at least 90% survival after 1040 hours," it means that the *Ratio of Improvement* (*x* axis of Figure 8.15) is 1040/1000 = 1.04. Notice that for the *Ratio of Improvement* of 1.04, the *Percent* (*y* axis of Figure 8.15), which represents the "power" of the test, for 2 maximum allowed failures is approximately 20% and that for 25 maximum allowed failures is approximately 80%. In other words, the test with 25 maximum allowed failures is more reliable than the one with 2 maximum allowed failures. Note that the former is also more expensive.

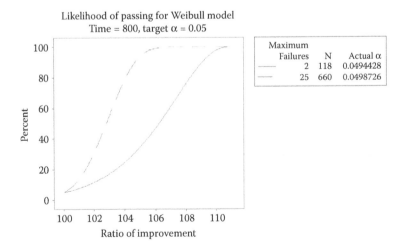

FIGURE 8.15
Step 7 in Section 8.2.

Notice also that as the *Ratio of Improvement* increases, the maximum allowed number of failures becomes less relevant. For example, if the true (unknown) performance of the motors is a lot higher than the standard that the test is aiming to show, the "power" of the test for 2 maximum allowed failures and that for 25 maximum allowed failures are the same (100%). In a way, the "power" of a test can be considered the likelihood of the management passing the test (conducted by the reliability engineer, in this case). Hence, the graph in Figure 8.15 is titled *Likelihood of Passing for Weibull Model*.

Another point to note is that as the *Ratio of Improvement* approaches 1, the maximum allowed number of failures becomes less relevant even then. For example, if the true (unknown) performance of the motors is equal to the standard that the test is aiming to show, the "power" of the test for 2 maximum allowed failures and that for 25 maximum allowed failures are the same.

We must add here that many reliability engineers are not concerned about the "power" of a test because the true performance of the products is unknown anyway.

8.3 Conclusion

This case study illustrated how to determine the test sample size using Minitab, when the test time and the number of failures are constrained. It also compared the outputs for different sets of constraints.

9

Reliability Test Plan with Constrained Sample Size and Number of Failures

Reliability tests are often conducted on a sample of products instead of the entire population, to save time and money. Many a time, it is prohibitively expensive or time consuming or both to gather a large sample of products for the test. Also, the company may not be willing to allow a large number of failures in the test, especially if the product is of high value. Amid such constraints, the reliability engineer is responsible for determining the testing time for the available sample. This case study illustrates how to determine the testing time using Minitab®, when the sample size and the number of failures are constrained.

A reliability engineer at a bearing manufacturing firm wishes to calculate the testing time that is required to assess the reliability of a new bearing, which is about to be launched in the market. The historical reliability standard for bearings produced by this firm has been "no more than 5% failures in one year of continuous use, which is 8760 hours." The firm has conducted prior studies on similar bearings, and believes that the "time to failure" distribution for the new bearing is Weibull, with a shape parameter of 3.

Given the cost constraints, the reliability engineer can afford to take a test sample of 20 bearings, and wants the experiment to result in no more than 3 failures. The engineer wishes to know how long she must test the sample, in order to claim that the new bearing satisfies the historical standard of "no more than 5% failures in 8760 hours."

Section 9.1 illustrates the approach to determine the testing time using Minitab for the historical standard of "no more than 5% failures in 8760 hours," with a test sample of 20 bearings and a maximum allowed number of failures of 3. Section 9.2 gives a few additional examples for different constraints and also compares the respective outputs. Finally, Section 9.3 gives some concluding remarks.

9.1 Determination of Testing Time

Perform the following steps to determine the testing time:

  Step 1: Open a new Minitab project file and go to *Demonstration*, as shown in Figure 9.1.

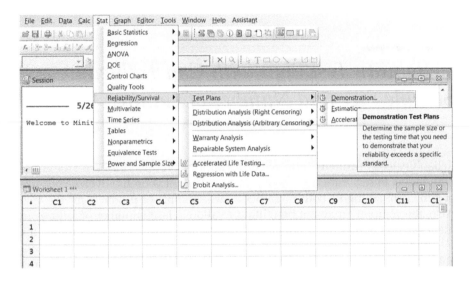

FIGURE 9.1
Step 1 in Section 9.1.

Step 2: The dialog box shown in Figure 9.2 opens.

Step 3: As shown in Figure 9.3, enter

- "8760" for *Percentile*
- "5" for *Percent*
- "3" for *Maximum number of failures allowed*
- "20" for *Sample sizes*
- "Weibull" for *Distribution*
- "3" for *Shape (Weibull) or scale (other distributions)*

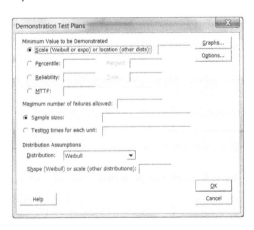

FIGURE 9.2
Step 2 in Section 9.1.

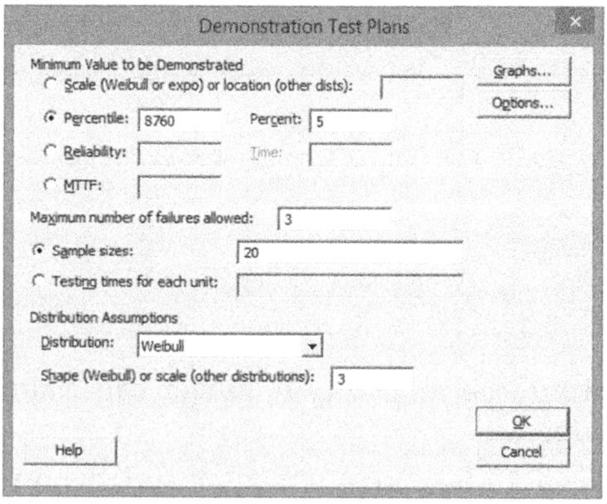

FIGURE 9.3
Step 3 in Section 9.1.

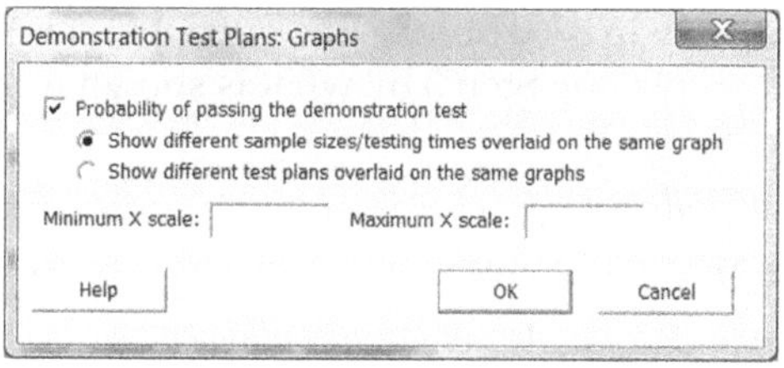

FIGURE 9.4
Step 4 in Section 9.1.

Step 4: Click on *Graphs*, and the dialog box shown in Figure 9.4 opens.

Step 5: As shown in Figure 9.5, uncheck the box for *Probability of passing the demonstration test.* (We are unchecking this box because it does not help to address the engineer's purpose in this case.)

Step 6: Click on *OK*, and it will take you back to the dialog box shown in Figure 9.3.

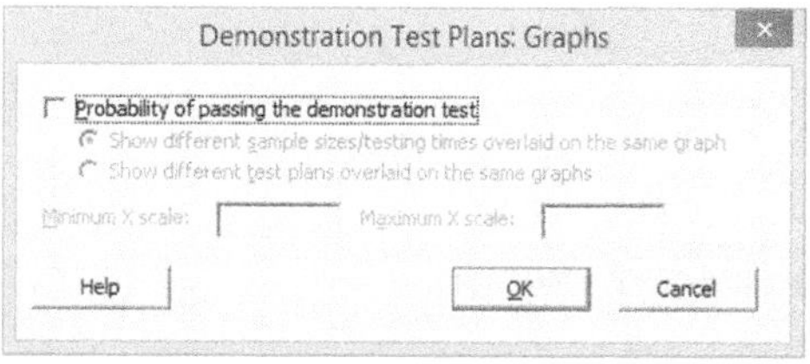

FIGURE 9.5
Step 5 in Section 9.1.

Step 7: Click on *OK*, and you will see the following output:

```
Reliability Test Plan
Distribution: Weibull, Shape = 3
Percentile Goal = 8760,   Actual Confidence Level = 95%


Failure  Sample  Testing
   Test    Size     Time
      3      20   17671.5
```

From this output, the reliability engineer can claim the following with 95% confidence:

If a sample of 20 bearings is tested and if no more than 3 of them fail in 17,671.5 hours, it means that the historical standard of "no more than 5% failures in 8760 hours" is achieved for all bearings.

Since "no more than 5% failures in 8760 hours" also means "at least 95% reliability in 8760 hours," the output that is shown can also be obtained by replacing Step 3 by what is shown in Figure 9.6. That is, enter

- "0.95" for *Reliability*
- "8760" for *Time*
- "3" for *Maximum number of failures allowed*
- "20" for *Sample sizes*
- "Weibull" for *Distribution*
- "3" for *Shape (Weibull) or scale (other distributions)*

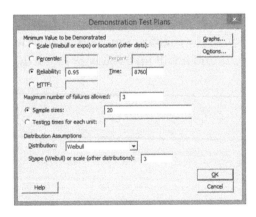

FIGURE 9.6
Alternative in Section 9.1.

9.2 Additional Scenarios

The reliability engineer now wishes to know what the testing time would be if she can afford to collect a larger sample size, for the same number of maximum allowed failures, that is, 3. To this end, she replaces Step 3 in Section 9.1 by performing what is shown in Figure 9.7. That is, enter

- "8760" for *Percentile*
- "5" for *Percent*
- "3" for *Maximum number of failures allowed*
- "40 60 80" for *Sample sizes*
- "Weibull" for *Distribution*
- "3" for *Shape (Weibull) or scale (other distributions)*

Notice from Figure 9.7 that the test is for three different sample sizes: 40 bearings, 60 bearings, and 80 bearings. The output is as follows:

```
Reliability Test Plan
Distribution: Weibull, Shape = 3
Reliability Goal = 0.95,   Actual Confidence Level = 95%


Failure   Sample   Testing
   Test     Size      Time
      3       40   13824.7
      3       60   12022.7
      3       80   10899.3
```

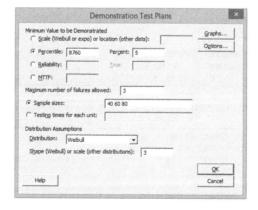

FIGURE 9.7
Test plan for different sample sizes in Section 9.2.

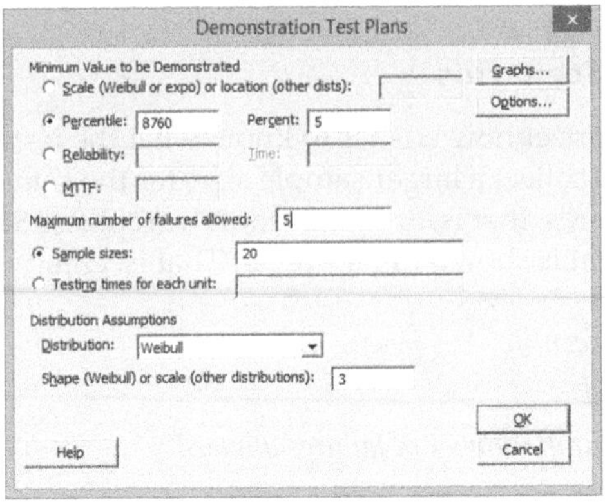

FIGURE 9.8
Test plan for five allowed failures in Section 9.2.

It is interesting to note from this output that for a fixed number of maximum allowed failures, as the sample size increases, the required testing time reduces. In other words, the larger the sample size is, the shorter the required testing time. For example, the testing time reduces from 13,824.7 hours to 12,022.7 hours, as the sample size increases from 40 bearings to 60 bearings.

The reliability engineer then wishes to know what the testing time would be if she can afford more than 3 failures, for the same sample size of 20 bearings. To this end, she replaces Step 3 in Section 9.1 by performing what is shown in Figures 9.8 through 9.10, one at a time.

The output for five allowed failures (Figure 9.8) is as follows:

```
Reliability Test Plan
Distribution: Weibull, Shape = 3
Reliability Goal = 0.95,   Actual Confidence Level = 95%



Failure  Sample  Testing
   Test    Size     Time
      5      20  19973.8
```

The output for 10 allowed failures (Figure 9.9) is as follows:

```
Reliability Test Plan
Distribution: Weibull, Shape = 3
Reliability Goal = 0.95,   Actual Confidence Level = 95%



Failure  Sample  Testing
   Test    Size     Time
     10      20  25036.5
```

The output for 15 allowed failures (Figure 9.10) is as follows:

```
Reliability Test Plan
Distribution: Weibull, Shape = 3
Reliability Goal = 0.95,  Actual Confidence Level = 95%


Failure  Sample  Testing
   Test    Size     Time
     15      20  30951.8
```

It is interesting to note from these outputs that for a fixed sample size, as the maximum allowed number of failures increases, the required testing time increases as well. In other words, to accommodate for more (potential) failures, the engineer is required to test the bearings for a longer period. For example, the testing time increases from 25,036.5 hours to 30,951.8 hours, as the maximum allowed number of failures increases from 10 failures to

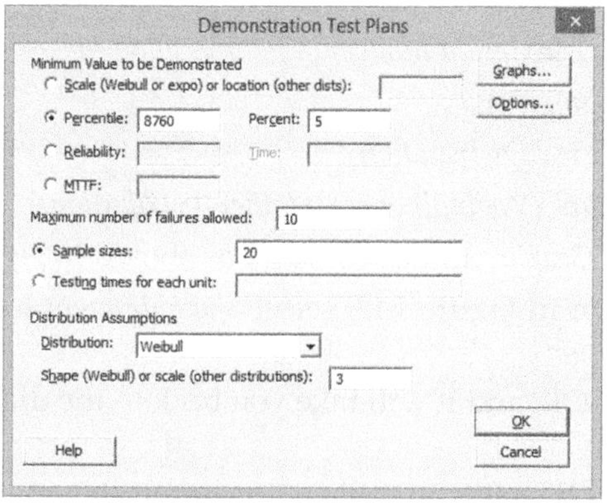

FIGURE 9.9
Test plan for 10 allowed failures in Section 9.2.

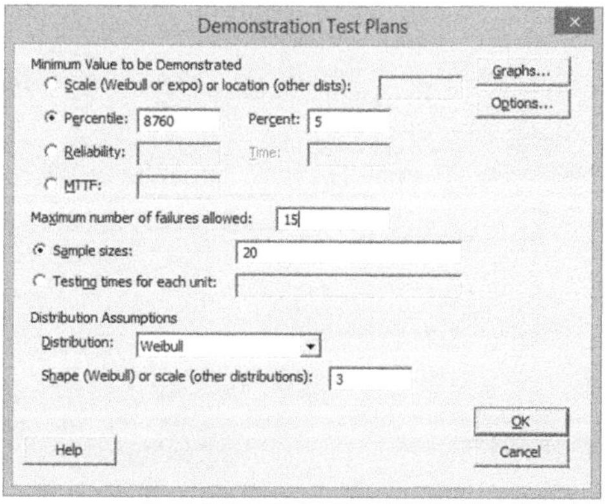

FIGURE 9.10
Test plan for 15 allowed failures in Section 9.2.

15 failures. An interesting question for the reader to think about is: What is the engineer's motivation for increasing the maximum allowed number of failures if it requires her to increase the testing time and hence increase the cost. The answer lies in the "power" of the demonstration test. The "power" gives us an indication of how reliable the test is. Suppose that the engineer wishes to compare the power of the tests for 1 maximum allowed failure and 15 maximum allowed failures. Perform the following steps:

Step 1: Open a new Minitab project file and go to *Demonstration*, as shown in Figure 9.1.

Step 2: The dialog box shown in Figure 9.2 opens.

Step 3: As shown in Figure 9.11, enter

- "8760" for *Percentile*
- "5" for *Percent*
- "1, 15" for *Maximum number of failures allowed*
- "20" for *Sample sizes*
- "Weibull" for *Distribution*
- "3" for *Shape (Weibull) or scale (other distributions)*

Step 4: Click on *Graphs*, and the dialog box shown in Figure 9.4 opens.

Step 5: As shown in Figure 9.12, select *Show different test plans overlaid on the same graphs*.

Step 6: Click on *OK*, and it will take you back to the dialog box shown in Figure 9.11.

Step 7: Click on *OK*, and you will see the graph shown in Figure 9.13.

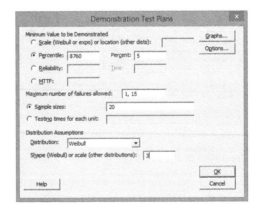

FIGURE 9.11
Step 3 in Section 9.2.

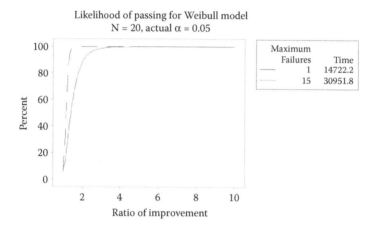

FIGURE 9.12
Step 5 in Section 9.2.

Likelihood of passing for Weibull model
N = 20, actual α = 0.05

Maximum Failures	Time
1	14722.2
15	30951.8

FIGURE 9.13
Step 7 in Section 9.2.

Recall that the reliability standard in this case study is "no more than 5% failures in 8760 hours." If the true (unknown) performance of the bearings is, say, "no more than 5% failures in 13,140 hours," it means that the *Ratio of Improvement* (x axis of Figure 9.13) is 13,140/8,760 = 1.50. Notice that for the *Ratio of Improvement* of 1.50, the *Percent* (y axis of Figure 9.13), which represents the "power" of the demonstration test, for 1 maximum allowed failures is approximately 50% and that for 15 maximum allowed failures is approximately 100%. In other words, the test with 15 maximum allowed failures is more reliable than the one with 1 maximum allowed failure. Note that the former is also more expensive.

Notice also that as the *Ratio of Improvement* increases, the maximum allowed number of failures becomes less relevant. For example, if the true (unknown) performance of the bearings is a lot higher than the standard that the test is aiming to show, the "power" of the demonstration test for 1 maximum allowed failure and that for 15 maximum allowed failures are the same (100%). In a way, the "power" of a test can be considered the likelihood of the management passing the test (conducted by the reliability engineer, in this case). Hence, the graph in Figure 9.13 is titled *Likelihood of Passing for Weibull Model*.

Another point to note is that as the *Ratio of Improvement* approaches 1, the maximum allowed number of failures becomes less relevant even then. For example, if the true (unknown) performance of the bearings is equal to the standard that the test is aiming to show, the "power" of the demonstration test for 1 maximum allowed failure and that for 15 maximum allowed failures are the same.

We must add here that many reliability engineers are not concerned about the "power" of the test because the true performance of the products is unknown anyway.

9.3 Conclusion

This case study illustrated how to determine the testing time using Minitab, when the test sample size and the number of failures are constrained. It also compared the outputs for different sets of constraints.

10

Comparison of Reliability of Parts from Different Vendors

Often, manufacturers use multiple vendors for parts, but find it difficult to compare the reliabilities of those parts. The *Regression with Life Data* option in Minitab® helps in such a situation.

In this case study, a reliability engineer wishes to test parts from two different vendors.

Section 10.1 discusses the failure time data from the vendors. Section 10.2 illustrates the regression analysis using Minitab. Section 10.3 gives an additional example using three different vendors. Finally, Section 10.4 gives a concluding remark.

10.1 Failure Time Data

The engineer tests 30 parts each from two different vendors: P and Q. The sample from Vendor P is tested for 150 hours and the sample from Vendor Q is tested for 100 hours.

Failure times (in hours) of 13 parts from Vendor P, which have failed, are recorded as follows: 7.5, 30.3, 38.4, 63.4, 71.7, 73.7, 84.7, 90.8, 92.9, 98.4, 100.9, 127.9, and 139.

Failure times (in hours) of 16 parts from Vendor Q, which have failed, are recorded as follows: 27.9, 27.2, 28.4, 30.8, 31.9, 33.3, 34.1, 53.6, 55.1, 61.6, 64.6, 65.1, 65.9, 68.7, 93.4, and 97.8.

Recall from Chapter 1 that a right-censored failure time means that the product that is being tested has not failed by that time. For example, if a product has not failed in, say, 100 hours, and if the test duration has ended, the failure time is recorded as 100 hours (right-censored). This product has not failed in 100 hours, but is expected to fail anytime later. Recall also that an exact failure time means that the product has failed exactly at that time. For example, if a product has failed exactly at the end of, say, 50 hours, the failure time is recorded as 50 hours (exact).

Failure times of 17 parts from Vendor P, which have not failed in 150 hours, are recorded as 150 hours (right-censored). Failure times of 14 parts from

Vendor Q, which have not failed in 100 hours, are recorded as 100 hours (right-censored).

10.2 Regression Analysis

Recall from Chapters 3 and 4 that it is possible to determine the statistical distribution that is the best fit for the failure time data. In this case study, we will assume that a lognormal distribution is the best fit for the failure time data from both vendors, and that the distribution parameter (scale) is the same for both vendors.

Perform the following steps for regression analysis:

Step 1: Open a new Minitab project file and enter the data, as shown in Figure 10.1. In the "Censor" column of the worksheet, "0" means right-censored failure time and "1" means exact failure time. The "Frequency" column shows the number of parts for which the failure time is applicable. For example, since there are 17 parts from Vendor P, whose failure times are recorded as 150 hours (right-censored), "17" is entered in the "Frequency" column for "150 hours."

Step 2: As shown in Figure 10.2, go to *Stat* → *Reliability/Survival* → *Regression with Life Data*. The dialog box shown in Figure 10.3 opens.

	File	Edit	Data	Calc	Stat	Graph	Editor
↓		C1	C2	C3		C4-T	
		Time	Censor	Frequency		Vendor	
1		7.5	1	1		P	
2		30.3	1	1		P	
3		38.4	1	1		P	
4		63.4	1	1		P	
5		71.7	1	1		P	
6		73.7	1	1		P	
7		84.7	1	1		P	
8		90.8	1	1		P	
9		92.9	1	1		P	
10		98.4	1	1		P	
11		100.9	1	1		P	
12		127.9	1	1		P	
13		139.0	1	1		P	
14		150.0	0	17		P	
15		27.9	1	1		Q	
16		27.2	1	1		Q	
17		28.4	1	1		Q	
18		30.8	1	1		Q	

FIGURE 10.1
Step 1 in Section 10.2.

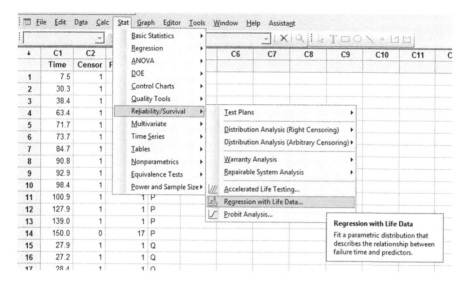

FIGURE 10.2
Step 2 commands in Section 10.2.

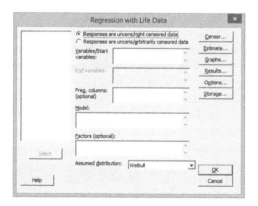

FIGURE 10.3
Step 2 dialog box in Section 10.2.

Step 3: As shown in Figure 10.4, select *Responses are uncens/right censored data*, select the "Time" column for *Variables/Start variables*, select the "Frequency" column for *Freq. columns*, select the "Vendor" column for *Model*, and select the "Vendor" column for *Factors*. Also, select "Lognormal" for *Assumed distribution*.

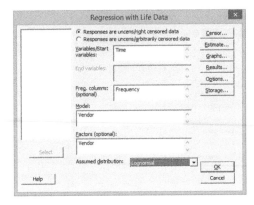

FIGURE 10.4
Step 3 in Section 10.2.

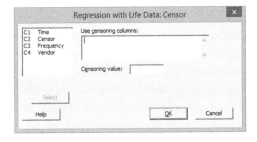

FIGURE 10.5
Step 4 in Section 10.2.

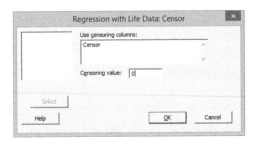

FIGURE 10.6
Step 5 in Section 10.2.

Step 4: Click on *Censor,* and the dialog box shown in Figure 10.5 opens.

Step 5: As shown in Figure 10.6, select the "Censor" column for *Use censoring columns* and enter "0" in *Censoring value.*

Step 6: Click on *OK,* and it will take you back to the dialog box shown in Figure 10.4.

Step 7: Click on *OK,* and you will see the following output in the session window:

```
Response Variable: Time
Frequency: Frequency

Censoring Information   Count
Uncensored value         29
Right censored value     31

Censoring value: Censor = 0

Estimation Method: Maximum Likelihood

Distribution:   Lognormal


Regression Table

                      Standard                    95.0% Normal CI
Predictor      Coef     Error      Z      P      Lower      Upper
Intercept   5.12391  0.223281  22.95  0.000   4.68628    5.56153
Vendor
Q          -0.543947  0.288012  -1.89  0.059  -1.10844   0.0205460
Scale       0.998827  0.146595                 0.749138   1.33174
```

We will assume here that the level of statistical significance is 0.07.

In this output, since the *p*-value (0.059) for Vendor Q is less than 0.07, there is sufficient evidence to conclude that failure time depends on vendor type. Also, since the coefficient (−0.543947) is negative for Vendor Q, Vendor Q is relatively worse than Vendor P.

10.3 Additional Example

The engineer tests 30 parts each from three different vendors: A, B, and C. The sample from Vendor A is tested for 150 hours, and the samples from Vendors B and C are tested for 100 hours.

Failure times (in hours) of 13 parts from Vendor A, which have failed, are recorded as follows: 7.5, 30.3, 38.4, 63.4, 71.7, 73.7, 84.7, 90.8, 92.9, 98.4, 100.9, 127.9, and 139.

Failure times (in hours) of 16 parts from Vendor B, which have failed, are recorded as follows: 27.9, 27.2, 28.4, 30.8, 31.9, 33.3, 34.1, 53.6, 55.1, 61.6, 64.6, 65.1, 65.9, 68.7, 93.4, and 97.8.

Failure times (in hours) of 16 parts from Vendor C, which have failed, are recorded as follows: 30, 29, 31, 33, 32, 24, 12, 34, 45, 32, 56, 57, 45, 43, 67, and 98.

Failure times of 17 parts from Vendor A, which have not failed in 150 hours, are recorded as 150 hours (right-censored). Failure times of 14 parts from Vendor B, which have not failed in 100 hours, are recorded as 100 hours (right-censored). Failure times of 14 parts from Vendor C, which have not failed in 100 hours, are recorded as 100 hours (right-censored).

In this case study, we will assume that a lognormal distribution is the best fit for the failure time data from all three vendors, and that the distribution parameter (scale) is the same for all three vendors.

Perform the following steps for regression analysis:

Step 1: Open a new Minitab project file and enter the data, as shown in Figure 10.7. In the "Censor" column of the worksheet, "0" means right-censored failure time and "1" means exact failure time. The "Frequency" column shows the number of parts for which the failure time is applicable. For example, since there are 17 parts from Vendor A, whose failure times are recorded as 150 hours (right-censored), "17" is entered in the "Frequency" column for "150 hours."

Step 2: As shown in Figure 10.8, go to *Stat* → *Reliability/Survival* → *Regression with Life Data*. The dialog box shown in Figure 10.9 opens.

File Edit Data Calc Stat Graph Editor			
↓	C1	C2	C3	C4-T
	Time	Censor	Frequency	Vendor
1	7.5	1	1	A
2	30.3	1	1	A
3	38.4	1	1	A
4	63.4	1	1	A
5	71.7	1	1	A
6	73.7	1	1	A
7	84.7	1	1	A
8	90.8	1	1	A
9	92.9	1	1	A
10	98.4	1	1	A
11	100.9	1	1	A
12	127.9	1	1	A
13	139.0	1	1	A
14	150.0	0	17	A
15	27.9	1	1	B
16	27.2	1	1	B
17	28.4	1	1	B

FIGURE 10.7
Step 1 in Section 10.3.

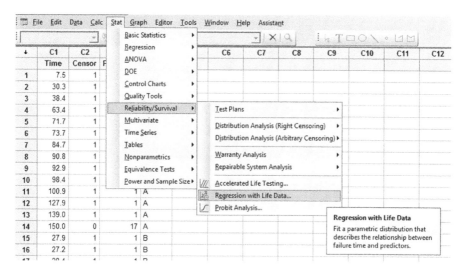

FIGURE 10.8
Step 2 commands in Section 10.3.

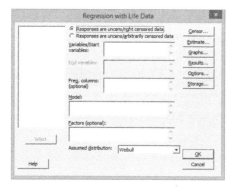

FIGURE 10.9
Step 2 dialog box in Section 10.3.

Step 3: As shown in Figure 10.10, select *Responses are uncens/right cen-sored data*, select the "Time" column for *Variables/Start variables*, select the "Frequency" column for *Freq. columns*, select the "Vendor" column for *Model*, and select the "Vendor" column for *Factors*. Also, select "Lognormal" for *Assumed distribution*.

Step 4: Click on *Censor*, and the dialog box shown in Figure 10.11 opens.

Step 5: As shown in Figure 10.12, select the "Censor" column for *Use censoring columns* and enter "0" in *Censoring value*.

Step 6: Click on *OK*, and it will take you back to the dialog box shown in Figure 10.10.

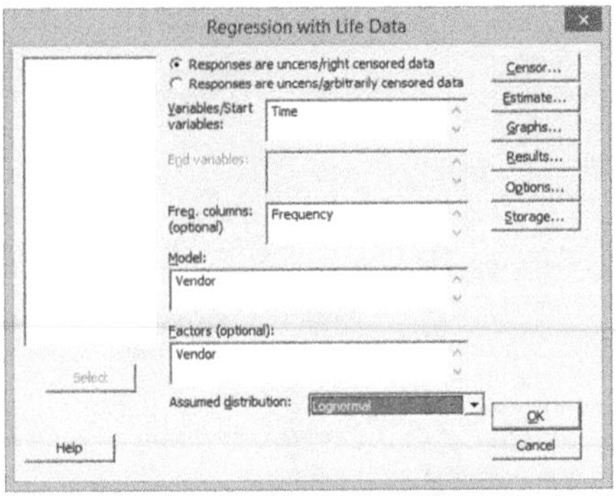

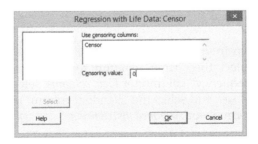

FIGURE 10.10
Step 3 in Section 10.3.

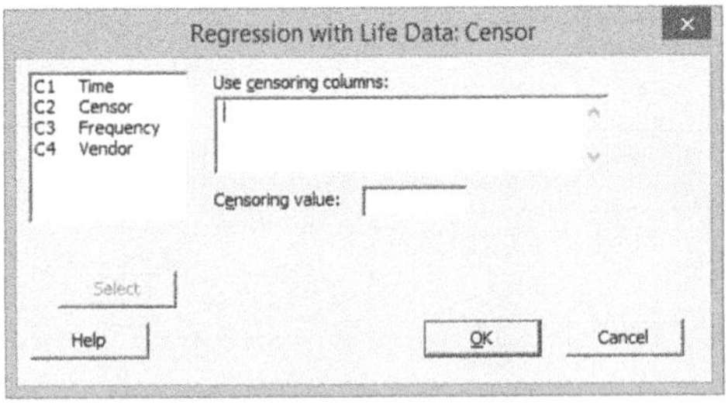

FIGURE 10.11
Step 4 in Section 10.3.

FIGURE 10.12
Step 5 in Section 10.3.

Step 7: Click on *OK*, and you will see the following output in the session window:

```
Response Variable: Time
Frequency: Frequency

Censoring Information   Count
Uncensored value          45
Right censored value      45

Censoring value: Censor = 0

Estimation Method: Maximum Likelihood

Distribution:   Lognormal


Regression Table

                        Standard                       95.0% Normal CI
Predictor        Coef      Error       Z       P      Lower      Upper
Intercept     5.12339   0.217340   23.57   0.000    4.69741    5.54937
Vendor
  B           -0.543829  0.287518   -1.89   0.059   -1.10735   0.0196954
  C           -0.696284  0.286376   -2.43   0.015   -1.25757   -0.134998
Scale          0.997934  0.117966                    0.791554    1.25812
```

We will assume here that the level of statistical significance is 0.07.

In this output

- Since the *p*-value (0.059) for Vendor B is less than 0.07, there is sufficient evidence to conclude that failure time depends on whether the vendor is A or B. Also, since the coefficient (−0.543829) is negative for Vendor B, Vendor B is relatively worse than Vendor A.

- Since the *p*-value (0.015) for Vendor C is less than 0.07, there is sufficient evidence to conclude that failure time depends on whether the vendor is A or C. Also, since the coefficient (−0.696284) is negative for Vendor C, Vendor C is relatively worse than Vendor A.

10.4 Conclusion

These case studies illustrated how to use the *Regression with Life Data* option in Minitab to compare the reliabilities of parts from different vendors.

11

Determination of Factors That Affect Product Reliability

Minitab®'s *Regression with Life Data* option that was introduced in Chapter 10 may also be used to identify more than one factor that may affect the life of a product.

In this case study, a reliability engineer at an auto parts manufacturer wishes to test whether the reliability of a car bumper depends on its weight and the machine that makes it.

Section 11.1 discusses the failure time data for car bumpers of three different weights and from two different machines. Section 11.2 illustrates the analysis using Minitab. Section 11.3 gives an additional example with four different weights and three different machines. Finally, Section 11.3 gives a concluding remark.

11.1 Failure Time Data

The engineer tests 48 randomly selected bumpers that weigh 5 lb., 7.5 lb., or 10 lb., and are made by either Machine A or Machine B. The failure times (in hours) are recorded as shown in Table 11.1. In the "Censor" column of the table, "Censored" means right-censored failure time and "Exact" means exact failure time. Recall from Chapter 1 that a right-censored failure time means that the product that is being tested has not failed by that time. For example, if a product has not failed in, say, 100 hours, and if the test duration has ended, the failure time is recorded as 100 hours (right-censored). This product has not failed in 100 hours, but is expected to fail anytime later. Recall also that an exact failure time means that the product has failed exactly at that time. For example, if a product has failed exactly at the end of, say, 50 hours, the failure time is recorded as 50 hours (exact).

TABLE 11.1

Failure Time Data for Section 11.2

No.	Failure	Censor	Machine	Weight
1	482	Censored	A	5.0
2	482	Censored	A	5.0
3	482	Censored	A	5.0
4	482	Censored	A	5.0
5	482	Censored	A	5.0
6	482	Censored	A	5.0
7	240	Exact	A	10.0
8	324	Exact	A	10.0
9	372	Exact	A	10.0
10	444	Exact	A	10.0
11	456	Exact	A	10.0
12	468	Exact	A	10.0
13	480	Censored	A	10.0
14	480	Censored	A	10.0
15	180	Exact	A	7.5
16	264	Exact	A	7.5
17	384	Exact	A	7.5
18	480	Censored	A	7.5
19	480	Censored	A	7.5
20	480	Censored	A	7.5
21	480	Censored	A	7.5
22	480	Censored	A	7.5
23	348	Exact	A	5.0
24	468	Exact	A	5.0
25	144	Exact	B	10.0
26	156	Exact	B	10.0
27	168	Exact	B	10.0
28	180	Exact	B	10.0
29	180	Exact	B	10.0
30	216	Exact	B	10.0
31	240	Censored	B	10.0
32	240	Censored	B	10.0
33	180	Exact	B	7.5
34	228	Exact	B	7.5
35	228	Exact	B	7.5
36	240	Censored	B	7.5
37	240	Censored	B	7.5
38	240	Censored	B	7.5

TABLE 11.1 (CONTINUED)

Failure Time Data for Section 11.2

No.	Failure	Censor	Machine	Weight
39	240	Censored	B	7.5
40	240	Censored	B	7.5
41	216	Exact	B	5.0
42	228	Exact	B	5.0
43	240	Censored	B	5.0
44	240	Censored	B	5.0
45	240	Censored	B	5.0
46	240	Censored	B	5.0
47	240	Censored	B	5.0
48	240	Censored	B	5.0

11.2 Determination of Factors Affecting Reliability

Recall from Chapters 3 and 4 that it is possible to determine the statistical distribution that is the best fit for the failure time data. In this case study, we will assume that a Weibull distribution is the best fit for the failure time data in Table 11.1.

Perform the following steps for the analysis:

Step 1: Open a new Minitab project file and enter the data from Table 11.1, as shown in Figure 11.1.

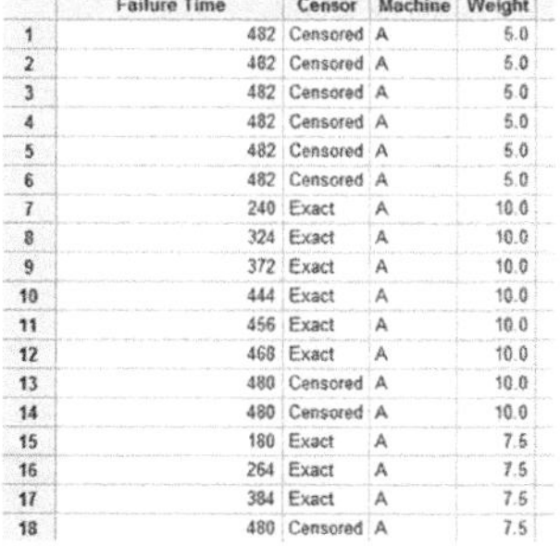

FIGURE 11.1
Step 1 in Section 11.2.

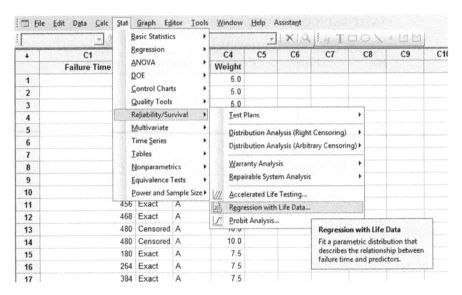

FIGURE 11.2
Step 2 commands in Section 11.2.

Step 2: As shown in Figure 11.2, go to *Stat* → *Reliability/Survival* → *Regression with Life Data*. The dialog box shown in Figure 11.3 opens.

Step 3: As shown in Figure 11.4, select *Responses are uncens/right censored data*, select the "Failure Time" column for *Variables/Start variables*, select the "Machine" and "Weight" columns for *Model*, and select the "Machine" and "Weight" columns for *Factors*. Also, select "Weibull" for *Assumed distribution*.

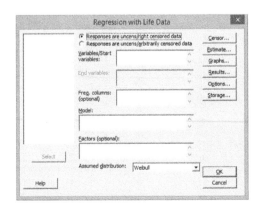

FIGURE 11.3
Step 2 dialog box in Section 11.2.

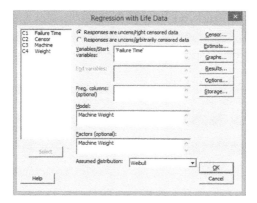

FIGURE 11.4
Step 3 in Section 11.2.

Step 4: Click on *Censor,* and the dialog box shown in Figure 11.5 opens.

Step 5: As shown in Figure 11.6, select the "Censor" column for *Use censoring columns* and enter "Censored" in *Censoring value.*

Step 6: Click on *OK,* and it will take you back to the dialog box shown in Figure 11.4.

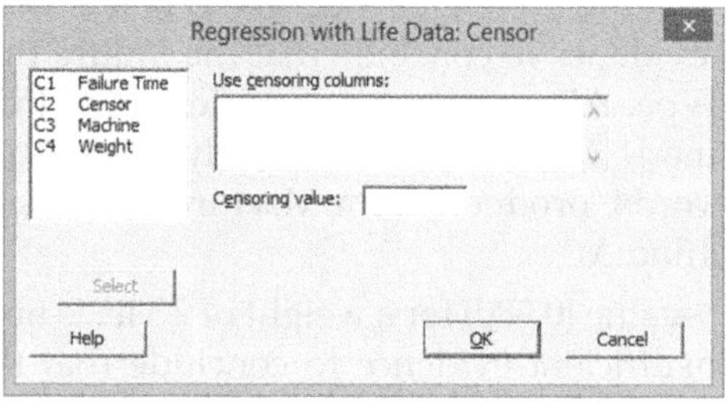

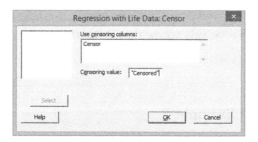

FIGURE 11.5
Step 4 in Section 11.2.

FIGURE 11.6
Step 5 in Section 11.2.

Step 7: Click on *OK*, and you will see the following output in the session window:

```
Response Variable: Failure Time

Censoring Information   Count
Uncensored value          22
Right censored value      26

Censoring value: Censor = Censored

Estimation Method: Maximum Likelihood

Distribution:   Weibull


Regression Table

                      Standard                      95.0% Normal CI
Predictor      Coef      Error      Z       P       Lower      Upper
Intercept   6.47812   0.136675   47.40   0.000    6.21025    6.74600
Machine
  B        -0.715589   0.0967284  -7.40   0.000   -0.905173  -0.526005
Weight
  7.5      -0.129372   0.147283   -0.88   0.380   -0.418040   0.159297
  10.0     -0.370072   0.142138   -2.60   0.009   -0.648658  -0.0914873
Shape      4.43928    0.854177                     3.04460    6.47284
```

We will assume here that the level of statistical significance is 0.05.

In this output

- Since the *p*-value (0.000) for Machine B is less than 0.05, there is sufficient evidence to conclude that the failure time depends on machine type. Also, since the coefficient (−0.715589) is negative for Machine B, Machine B is relatively worse than Machine A. In other words, products from Machine B fail faster than those from Machine A.

- Since the *p*-value (0.380) for a weight of 7.5 lb. is not less than 0.05, there is insufficient evidence to conclude that the failure time depends on whether the weight is 5 lb. or 7.5 lb.

- Since the *p*-value (0.009) for a weight of 10 lb. is less than 0.05, there is sufficient evidence to conclude that the failure time depends on whether the weight is 5 lb. or 10 lb. Also, since the coefficient (−0.370072) is negative for a weight of 10 lb., a weight of 10 lb. is relatively worse than a weight of 5 lb.

11.3 Additional Example

The engineer tests 48 randomly selected bumpers that weigh 1 lb., 2 lb., 3 lb., or 4 lb., and are made by either Machine P or Machine Q or Machine R. The failure times (in hours) are recorded as shown in Table 11.2. In the "Censor"

TABLE 11.2

Failure Time Data for Section 11.3

No.	Failure	Censor	Machine	Weight
1	482	Censored	P	1
2	482	Censored	P	1
3	482	Censored	P	1
4	482	Censored	P	1
5	482	Censored	P	2
6	482	Censored	P	2
7	240	Exact	P	2
8	324	Exact	P	2
9	372	Exact	P	3
10	444	Exact	P	3
11	456	Exact	P	3
12	468	Exact	P	3
13	480	Censored	P	4
14	480	Censored	P	4
15	180	Exact	P	4
16	264	Exact	P	4
17	384	Exact	Q	1
18	480	Censored	Q	1
19	480	Censored	Q	1
20	480	Censored	Q	1
21	480	Censored	Q	2
22	480	Censored	Q	2
23	348	Exact	Q	2
24	468	Exact	Q	2
25	144	Exact	Q	3
26	156	Exact	Q	3
27	168	Exact	Q	3
28	180	Exact	Q	3
29	180	Exact	Q	4
30	216	Exact	Q	4
31	240	Censored	Q	4
32	240	Censored	Q	4
33	180	Exact	R	1
34	228	Exact	R	1
35	228	Exact	R	1
36	240	Censored	R	1
37	240	Censored	R	2
38	240	Censored	R	2
39	240	Censored	R	2
40	240	Censored	R	2

(Continued)

TABLE 11.2 (CONTINUED)

Failure Time Data for Section 11.3

No.	Failure	Censor	Machine	Weight
41	216	Exact	R	3
42	228	Exact	R	3
43	240	Censored	R	3
44	240	Censored	R	3
45	240	Censored	R	4
46	240	Censored	R	4
47	240	Censored	R	4
48	240	Censored	R	4

column of the table, "Censored" means right-censored failure time and "Exact" means exact failure time.

In this example, we will assume that a Weibull distribution is the best fit for the failure time data in Table 11.2.

Perform the following steps for the analysis:

Step 1: Open a new Minitab project file and enter the data from Table 11.2, as shown in Figure 11.7.

Step 2: As shown in Figure 11.8, go to *Stat → Reliability/Survival → Regression with Life Data*. The dialog box shown in Figure 11.9 opens.

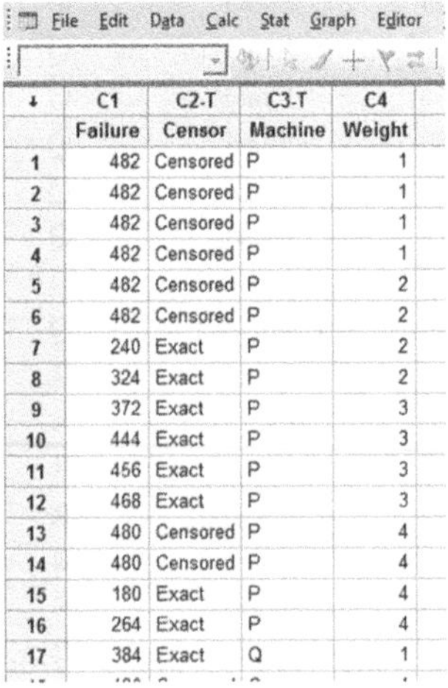

FIGURE 11.7
Step 1 in Section 11.3.

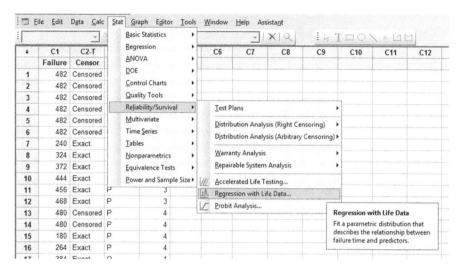

FIGURE 11.8
Step 2 commands in Section 11.3.

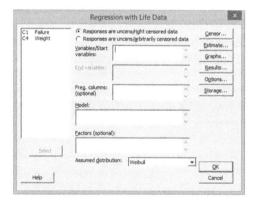

FIGURE 11.9
Step 2 dialog box in Section 11.3.

Step 3: As shown in Figure 11.10, select *Responses are uncens/right censored data*, select the "Failure" column for *Variables/Start variables*, select the "Machine" and "Weight" columns for *Model*, and select the "Machine" and "Weight" columns for *Factors*. Also, select "Weibull" for *Assumed distribution*.

Step 4: Click on *Censor*, and the dialog box shown in Figure 11.11 opens.

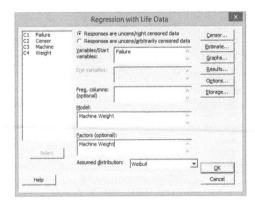

FIGURE 11.10
Step 3 in Section 11.3.

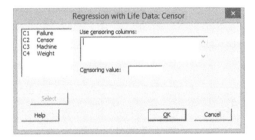

FIGURE 11.11
Step 4 in Section 11.3.

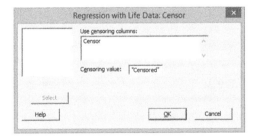

FIGURE 11.12
Step 5 in Section 11.3.

Step 5: As shown in Figure 11.12, select the "Censor" column for *Use censoring columns* and enter "Censored" in *Censoring value*.

Step 6: Click on *OK*, and it will take you back to the dialog box shown in Figure 11.10.

Step 7: Click on *OK*, and you will see the following output in the session window:

```
Response Variable: Failure

Censoring Information   Count
Uncensored value          22
Right censored value      26

Censoring value: Censor = Censored

Estimation Method: Maximum Likelihood

Distribution:   Weibull


Regression Table

                       Standard                    95.0% Normal CI
Predictor       Coef      Error     Z      P      Lower      Upper
Intercept    6.75663   0.221477  30.51  0.000    6.32254    7.19071
Machine
Q            -0.546287  0.186850  -2.92  0.003  -0.912506  -0.180068
R            -0.462630  0.184315  -2.51  0.012  -0.823882  -0.101379
Weight
2            -0.0313143 0.224627  -0.14  0.889  -0.471575   0.408946
3            -0.773909  0.222956  -3.47  0.001  -1.21090   -0.336923
4            -0.449067  0.238471  -1.88  0.060  -0.916462   0.0183286
Shape         3.15108   0.572350                 2.20724    4.49851
```

We will assume here that the level of statistical significance is 0.05.

In this output

- Since the *p*-value (0.003) for Machine Q is less than 0.05, there is sufficient evidence to conclude that the failure time depends on whether it is Machine P or Machine Q. Also, since the coefficient (−0.546287) is negative for Machine Q, Machine Q is relatively worse than Machine P. In other words, products from Machine Q fail faster than those from Machine P.

- Since the *p*-value (0.012) for Machine R is less than 0.05, there is sufficient evidence to conclude that the failure time depends on whether it is Machine P or Machine R. Also, since the coefficient (−0.462630) is negative for Machine R, Machine R is relatively worse than Machine P.

- Since the coefficient (−0.462630) for Machine R is higher than the coefficient (−0.546287) for Machine Q, Machine R is better than Machine Q.

- Since the *p*-value (0.889) for a weight of 2 lb. is not less than 0.05, there is insufficient evidence to conclude that the failure time depends on whether the weight is 1 lb. or 2 lb.

- Since the *p*-value (0.001) for a weight of 3 lb. is less than 0.05, there is sufficient evidence to conclude that the failure time depends on whether the weight is 1 lb. or 3 lb. Also, since the coefficient (−0.773909) is negative for a weight of 3 lb., the weight of 3 lb. is relatively worse than the weight of 1 lb.

- Since the *p*-value (0.060) for a weight of 4 lb. is not less than 0.05, there is insufficient evidence to conclude that the failure time depends on whether the weight is 1 lb. or 4 lb.

11.4 Conclusion

These case studies illustrated how to use the *Regression with Life Data* option in Minitab to test whether the reliability of a product depends on certain factors.

12

Prediction of Stress Levels That Cause Product Failure

Minitab®'s *Probit Analysis* option is useful in predicting the stress levels that cause the failure of a product, by considering the proportions of product failures at various stress levels. Readers interested in the mathematical formulas of probit analysis are encouraged to review the book listed in the Bibliography.

In this case study, a reliability engineer at a light bulb manufacturer wishes to predict the voltage level at which a light bulb will fail after a certain test period.

Section 12.1 illustrates how to perform probit analysis. Section 12.2 interprets the output of the analysis. Finally, Section 12.3 gives a concluding remark.

12.1 Probit Analysis

The engineer tests 250 randomly selected light bulbs in batches of 50 each, with different voltage levels as shown in Table 12.1. Also, as shown in the table, the number of failed bulbs in each batch is recorded at the end of 700 hours of test time. We will assume that a Weibull distribution is the best fit for the failure times of light bulbs.

The engineer wishes to know the answers to the following questions:

- At what voltage are 75% of light bulbs expected to fail in 700 hours (test time)?
- What percentage of light bulbs are expected to survive past 700 hours (test time) at 118 volts?

Perform the following steps for the analysis:

Step 1: Open a new Minitab project file and enter the data from Table 12.1, as shown in Figure 12.1.

Step 2: As shown in Figure 12.2, go to *Stat → Reliability/Survival → Probit Analysis*. The dialog box shown in Figure 12.3 opens.

TABLE 12.1

Test Results

No. of Failures	No. of Test Bulbs	Voltage
2	50	107
6	50	114
11	50	120
27	50	126
45	50	132

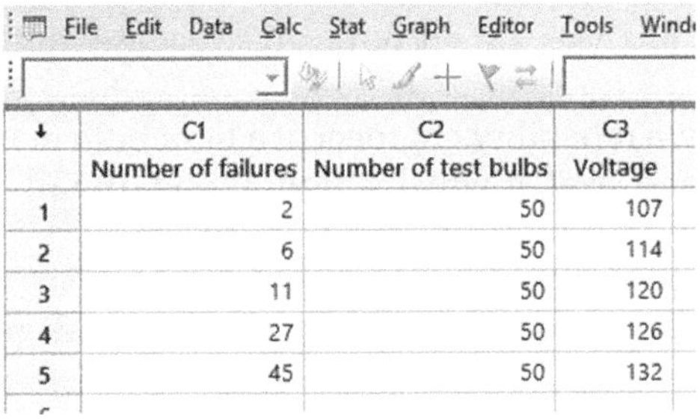

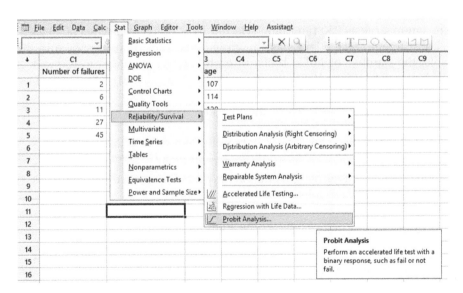

FIGURE 12.1
Step 1 in Section 12.1.

FIGURE 12.2
Step 2 commands in Section 12.1.

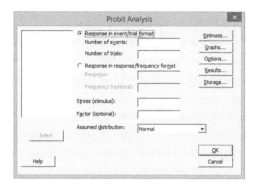

FIGURE 12.3
Step 2 dialog box in Section 12.1.

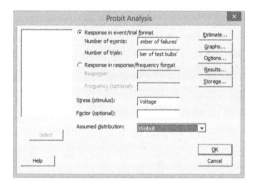

FIGURE 12.4
Step 3 in Section 12.1.

Step 3: As shown in Figure 12.4, select the "Number of failures" column for *Number of events*, select the "Number of test bulbs" column for *Number of trials*, and select the "Voltage" column for *Stress (stimulus)*. Also, select "Weibull" for *Assumed distribution*.

Step 4: Click on *Estimate*, and the dialog box shown in Figure 12.5 opens.

Step 5: As shown in Figure 12.6, enter "75" for *Estimate percentiles for these additional percents* and enter "118" for *Estimate probabilities for these stress values*.

Step 6: Click on *OK*, and it will take you back to the dialog box shown in Figure 12.4.

Step 7: Click on *Graphs*, and the dialog box shown in Figure 12.7 opens.

Step 8: As shown in Figure 12.8, uncheck the boxes for *Probability plot* and *Display confidence intervals on above plots*. Also, check the box for *Survival plot*.

Step 9: Click on *OK*, and it will take you back to the dialog box shown in Figure 12.4.

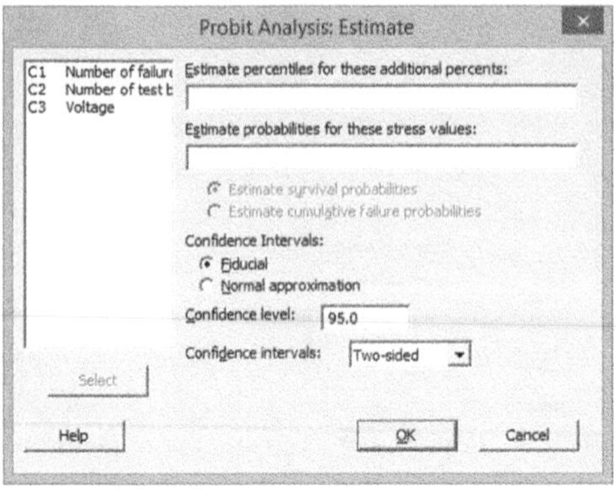

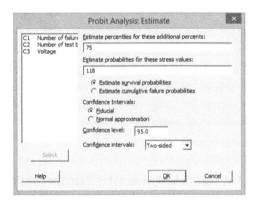

FIGURE 12.5
Step 4 in Section 12.1.

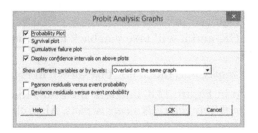

FIGURE 12.6
Step 5 in Section 12.1.

FIGURE 12.7
Step 7 in Section 12.1.

FIGURE 12.8
Step 8 in Section 12.1.

Step 10: Click on *OK*, and you will see the following output in the session window:

```
Goodness-of-Fit Tests

Method      Chi-Square  DF      P
Pearson        1.14854   3   0.765
Deviance       1.12513   3   0.771


Table of Percentiles

                                      95.0% Fiducial
                            Standard        CI
Percent  Percentile      Error    Lower    Upper
      1     101.754    2.54695   95.5547  105.939
      2     105.261    2.23934   99.7966  108.938
      3     107.379    2.04954  102.374   110.745
      4     108.916    1.91047  104.250   112.056
      5     110.130    1.80013  105.734   113.091
      6     111.137    1.70842  106.968   113.950
      7     112.000    1.62983  108.026   114.687
      8     112.757    1.56104  108.955   115.334
      9     113.433    1.49984  109.784   115.913
     10     114.045    1.44472  110.534   116.437
     20     118.274    1.07859  115.704   120.099
     30     120.997    0.876081 118.976   122.530
     40     123.125    0.759280 121.453   124.518
     50     124.962    0.707231 123.494   126.340
     60     126.666    0.711579 125.279   128.143
     70     128.356    0.767270 126.942   130.042
     75     129.237    0.814121 127.773   131.071
     80     130.177    0.875149 128.637   132.192
     90     132.459    1.05977  130.662   134.994
     91     132.747    1.08590  130.912   135.353
     92     133.055    1.11443  131.178   135.739
     93     133.388    1.14590  131.465   136.158
     94     133.753    1.18111  131.778   136.619
     95     134.161    1.22126  132.127   137.136
     96     134.630    1.26831  132.525   137.731
     97     135.190    1.32582  132.999   138.446
     98     135.910    1.40145  133.604   139.368
     99     136.990    1.51824  134.506   140.760


Table of Survival Probabilities

                           95.0% Fiducial CI
Stress  Probability      Lower      Upper
   118      0.808389   0.745640   0.874853
```

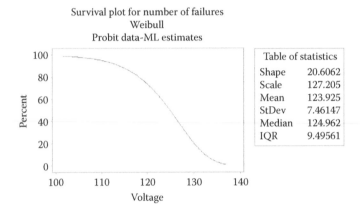

FIGURE 12.9
Survival plot in Section 12.2.

12.2 Interpretation of Output

The following is clear from the output at the end of Section 12.1:

- Since the *p*-values (0.765 and 0.771) for both Pearson and Deviance methods are large (at least compared with the typical lower limit of 0.70), there is sufficient evidence to conclude that the failure times of the light bulbs do follow Weibull distribution.
- 75% of light bulbs are expected to fail in 700 hours (test time) at 129.237 volts.
- 80.8389% of light bulbs are expected to survive past 700 hours (test time) at 118 volts.

In addition to the output in the session window, the survival plot shown in Figure 12.9 is a result of the analysis. It is evident from the plot that about 20% of light bulbs are expected to survive past 700 hours (test time) at about 130 volts.

12.3 Conclusion

This case study illustrated how to use the *Probit Analysis* option in Minitab to predict the stress level at which a product will fail after a certain test period.

Bibliography

Elsayed, E. A. *Reliability Engineering*, 2nd Edn, New York: Wiley, 2012.

Fox, J. *Applied Regression Analysis and Generalized Linear Models*, 2nd Edn, New York: Sage, 2008.

Kapur, K. C., and Pecht, M. *Reliability Engineering* (Wiley Series in Systems Engineering and Management), 1st Edn, New York: Wiley, 2014.

Krishnamoorthy, K. *Handbook of Statistical Distributions with Applications* (Statistics: A Series of Textbooks and Monographs), Boca Raton, FL: Chapman and Hall/CRC, 2006.

Nelson, W. B. *Accelerated Testing: Statistical Models, Test Plans, and Data Analysis*, 1st Edn, New York: Wiley, 2004.

Index

Printed and bound by CPI Group (UK) Ltd, Croydon, CR0 4YY

23/10/2024

01777697-0011